BLUEPRINTS FOR TROPICAL DAIRY FARMING

BLUEPRINTS FOR TROPICAL DAIRY FARMING

Increasing Domestic Milk Production in Developing Countries

John Moran and Philip Chamberlain

CSIRO
PUBLISHING

National Library of Australia Cataloguing-in-Publication entry

Moran, John, 1945– author.

Blueprints for tropical dairy farming : increasing domestic milk production in developing countries / John Moran and Philip Chamberlain.

9781486306466 (paperback)
9781486306473 (epdf)
9781486306480 (epub)

Includes bibliographical references and index.

Dairying – Southeast Asia.
Milk yield – Southeast Asia.
Dairy farming – Southeast Asia – Management.
Dairy farms – Southeast Asia – Management.

Chamberlain, Philip, author.

636.21420959

Published by

CSIRO Publishing
36 Gardiner Road, Clayton VIC 3168
Private Bag 10, Clayton South VIC 3169
Australia

Telephone: [+613] 9545 8555
Email: csiropublishing@csiro.au
Website: www.publishing.csiro.au

Front cover: Imported Australian Friesian cows resting in the dirt yard next to their free stall shed in north Vietnam. These cows average 20 to 24 L milk/cow/day.
Back cover: The best way to develop blueprints is within a group, so you can share your ideas. This photo shows dairy advisers from Thailand.

All photographs are by John Moran and Philip Chamberlain unless otherwise indicated.

Set in 11/13.5 Minion & Helvetica Neue
Edited by Anne Findlay
Cover design by James Kelly
Typeset by Thomson Digital
Printed by Ingram Lightning Source

Feb26_RP_ILS

Foreword

It gives me great honour and pleasure to write this Foreword to the latest of John Moran's books, namely *Blueprints for Tropical Dairy Farming.* I have known John for many years, when he was a dairy researcher and adviser for the Victorian government in Australia and following his retirement in 2011, as a tropical dairy consultant through his business Profitable Dairy Systems.

As a result of writing five reference texts on tropical dairy farming over the last 10 years, Dr John Moran has become 'the guru' for dairy advisers, policymakers, farmers and other dairy stakeholders throughout South and East Asia. This, his sixth book, follows on with his tradition of thoroughly reading, researching and networking his topics before putting 'pen to paper' (or its modern day equivalent: putting 'finger to keyboard'). There has been much recognition of his skills and credibility such as his recent three-year term as the Coordinator of the Asia Dairy Network, an FAO body facilitating the commercial developments of smallholder dairy farming throughout Asia. In addition, when you Google 'dairy farming in the humid tropics' you will invariably come across many of John's writings and webinars within the first few lines of the printout.

He has called his sixth book *Blueprints for Tropical Dairy Farming* because, in collaborating with his colleague, Dr Phillip Chamberlain, another Australian tropical livestock guru, John wants to cover all aspects of producing milk in the tropics, namely by including detailed discussions on large-scale intensive dairy feedlotting as well as the more traditional smallholder dairy systems. This relatively new approach in the tropics is being seen by many as the best and most appropriate way to address the ever-increasing gap between demand and supply of raw milk leading to decreasing self-sufficiencies of milk and dairy products in far too many tropical Asian countries.

Other very new innovations in John's kitbag are several smallholder dairy farmer extension packages, specifically developed to address one of the biggest problems in tropical dairy farming, namely 'getting the message across' from dairy researchers to teachers and trainers and eventually to the farmers themselves. In this book, John and Philip discuss two of these, namely 'The Golden Rules of tropical dairy farming' and the 'Dairy Smallholder Innovative Program (or DSIP)'. Their key objectives are to develop individual farmer skills, primarily in dairy production technology, to improve the productivity of their milking herds to stimulate farm performance and bring about profitable, hence long-term, sustainability.

It is not easy to write a book about tropical dairy farming in which every fact is relevant to every reader. However, John and Phillip have done an admirable job which will greatly add to our collective knowledge by developing the improved farm management practices that are essential to overcome many traditions in smallholder dairy farming.

Not only will this book be of great interest to dairy farmers and advisers throughout the tropics, but university and college students, dairy scientists and extension personnel will also find it a valuable reference text on tropical dairy farming. I thoroughly recommend this book to all.

Professor Metha Wanapat
Director, Tropical Feed Resources Research and Development Centre (TROFREC)
Faculty of Agriculture, Khon Kaen University
Khon Kaen, Thailand
Email: metha@kk.ac.th

Contents

About the authors

John Moran

For over 30 years Dr John Moran was an Australian senior research and advisory scientist from Victoria's Department of Primary Industries (DPIV), located in northern Victoria. Over the last 10 years he has spent half his time advising farmers in southern Australia and half his time working with dairy farmers and advisers in South and East Asia. His specialist fields include milk production, ruminant nutrition, calf and heifer rearing, forage conservation and whole farm business management. Following his retirement from DPIV in June 2011, John has managed his consulting business, Profitable Dairy Systems, located in Kyabram.

John graduated in 1967 with a Rural Science honours degree from New England University at Armidale in NSW, followed by a Masters degree in 1969. In 1976 he obtained a Doctorate of Philosophy in beef production from University of London, Wye College in England. During the 1980s, John lived in Indonesia for three years, working in beef cattle and buffalo research. Since 1999, he initiated and conducted training programs on smallholder dairy production to farmers, advisers and policymakers in Indonesia, Malaysia, Thailand, Vietnam, China, Pakistan, Sri Lanka, Bangladesh, Myanmar and East Timor. In 2013 he was appointed as the Coordinator of the Asia Dairy Network, an international agency position to facilitate communication and collaboration of dairy industries from 15 countries throughout South and East Asia. In addition to the smallholder dairy industries of tropical Asian countries, John is now consulting for large-scale dairy feedlot farms in Indonesia, Vietnam, Malaysia and India.

As a result of his Asian programs, in 2005 John wrote his first dairy manual, *Tropical Dairy Farming: Feeding Management for Smallholder Dairy Farmers in the Humid Tropics*. As well as hard copy, the book was published on the internet, making it freely available to dairy stakeholders throughout the world. In 2008 John wrote his second manual on tropical dairy farming, namely *Business Management for Tropical Dairy Farmers*. By June 2013, these two books had received over 120 000 'hits' on the internet, indicating their relevance particularly to Asian tertiary teaching institutes and government livestock departments.

The in-depth review he conducted during a LIVECORP consultancy on developing management packages for tropical dairy production technology formed the basis of his third book, entitled *Managing High Grade Dairy Cows in the Tropics*. His fourth book, entitled *Rearing Young Stock on Tropical Farms in Asia,* was an update of one of his earlier books, but written specifically for tropical smallholder farmers. His fifth book on tropical dairy farming, *Cow Talk: Understanding Dairy Cow Behaviour to Improve Their Welfare on Asian Farms* concentrates on animal welfare, a very valid topic for the current emphasis on the importation of exotic dairy heifers to many tropical countries. All his tropical dairy farming manuals are freely downloadable from www.publish.csiro.au.

John has published more than 200 research papers and advisory articles. He has also written several farmer manuals on dairy and beef cattle nutrition, veal production calf and heifer rearing and on silage production. All his books are commercially available through CSIRO Publishing. In 2016, John was invited to Kenya to develop a farmer manual for smallholder farmers in that country.

John's initial training in a systems approach to livestock science, together with his many years working closely with dairy industries in Australia and South and East Asia, stands him in great stead to write *Blueprints for Tropical Dairy Farming*, his sixth tropical manual, which complements his previous books. This book was written in conjunction with Dr Philip Chamberlain, who has collaborated with John on several recent consultancies.

Philip Chamberlain

Philip is an Australian veterinary and project management consultant, with over 25 years' experience in tropical dairy development, production, genetics, health, management and strategic planning in northern Australia, Asia and Europe. He teaches dairy and aquaculture husbandry and management at several international veterinary schools in the Asian region, as well as developing veterinary science and animal production curricula at universities in Australia, Vietnam and Cambodia. In addition, he provides advice and training programs to large and smallholder cattle farm developers in Malaysia, Myanmar, China, Indonesia and India, as well as investment and development advice to large-scale cattle operators in China. He is currently managing several large-scale dairy extension training programs in China and India, which include the importation of suitable tropical dairy genetics through *in vitro* fertilisation (IVF).

Other books and technical manuals written by the senior author

Books

Calf Rearing: A Guide to Rearing Calves in Australia (1993)
Forage Conservation: Making Quality Silage and Hay in Australia (1996)
Heifer Rearing: A Guide to Rearing Dairy Replacement Heifers in Australia (with Douglas McLean) (2001)
Calf Rearing: A Practical Guide (2002)
Tropical Dairy Farming: Feeding Management for Smallholder Dairy Farmers in the Humid Tropics (2005)
Business Management for Tropical Dairy Farmers (2009)
Feedpads for Grazing Dairy Cows (with Scott McDonald) (2010)
Fifty Years of Farmer Extension for Victoria's Dairy Industry (2011)
Managing High Grade Dairy Cows in the Tropics (2012)
Rearing Young Stock on Tropical Dairy Farms in Asia (2012)
Feeding Management of Smallholder Dairy Cattle and Buffalo in Tropical Asia: An e-learning Course (2013)
Cow Talk: Understanding Dairy Cow Behaviour to Improve Their Welfare on Asian Farms (with Rebecca Doyle) (2015)
Smallholder Dairy Farming in Kenya. The Golden Rules for Improved Cow Performance and Greater Farm Profits (2016)

Technical manuals

Maize for Fodder: A Guide to Growing, Conserving and Feeding Irrigated Maize in Northern Victoria (with Ken Pritchard) (1987)
Growing Calves for Pink Veal: A Guide to Rearing, Feeding and Managing Calves for Pink Veal in Australia (1990)
Growing Quality Forages for Smallholder Dairy Farms in Indonesia (2001)
Managing Dairy Farm Costs: Strategies for Dairy Farmers in Irrigated Northern Victoria (2002)
Feeding Management for Smallholder Dairy Farmers in Thailand (2002)
Improving Milk Composition Through Better Feeding Management (2003)

The Key Drivers of Good Reproductive Performance on Indonesian Dairy Farms (2006)
Value-adding Indonesia's Dairy Industry: Developing Cottage Industries in East Java (2006)
Improving Business Skills of Smallholder Dairy Farmers in Thailand (2007)
Dairy Production in Malaysia with Particular Reference to Milk Quality (2007)
Dairy Production in Indonesia with Particular Reference to Milk Quality (2007)
Developing a Post-arrival Herd Management Protocol for Imported Australian Dairy Heifers (2007)
Managing Heat Stress and Housing for Pakistani Dairy Cows and Buffaloes (2007)
Dairy Production from Smallholder Farmers in Asia: Synopses of Workshops from 2000 to 2007 (2008)
Dairy Production in Sri Lanka: Current Situation and Future Prospects (2008)
Improving Business Skills in Vietnam's Smallholder Dairy Industry (2008)
A Guide to Better Dairy Herd Management in the Tropics (2009)
Best Management Practices for Smallholder Dairy Farmers in Asia (2010)
Developing a Blueprint for a 50-cow Smallholder Dairy Farm in Asia (2010)
The Nutritive Value of Feedstuffs for Dairy Farmers in South-East Asia (2012)
A Blueprint for a Large-scale Intensive Dairy Farm in Asia (2012)
Developing Standard Operating Procedures for the Welfare of Australian Dairy Heifers in Overseas Markets (2013)
The Feeding of By-products on Smallholder Dairy Farms in Asia and Other Tropical Regions (2014)
Improving the Feeding and Herd Management on Intensively Managed Dairy Feedlots in North Vietnam (2016)
Improving the Feeding and Herd Management on Kenya's Smallholder Dairy Farms (2016)

Acknowledgements

For over 20 years we – John Moran and Philip Chamberlain – have been very fortunate to be able to travel around many countries in South and East Asia and (for John), more recently, in Africa. The reasons for our travels are many and varied, but they are primarily related to opportunities to learn more about how farmers can make a better living out of their dairy cows and to share that knowledge with other farmers, university students and other stakeholders in Asia's dairy industries. These travels have taught us so much about the tenacity of the human spirit because humans the world over love their cows and even love to tell others about them. Life as dairy farmer is a 24/7 devotion to their stock, whether they like it or not: generally dairy farmers do this each and every day because it can, and generally does, bring them a relatively quick return for their farming efforts. Cows require milking twice daily and because raw milk is so perishable, it must be quickly sold into the next step in the value chain, usually the milk trader, cooperative or milk-processing agent. It is little wonder that recent estimates of dairy farmers the world over (which includes people living within dairy farming households) are of the order of 750 to 900 million people.

There are so many people we must acknowledge and thank for their input into satisfying our lifelong thirst for knowledge. They have provided insights into the background technical information we have today, as well as the practical skills on 'how we can get our cows to produce more milk and more baby calves'. If we mention any names, it would be unfair to others whom we do not mention. Therefore, we will generally thank the thousands and thousands of dairy farmers from whom we have both learnt much, as well as the hundreds and hundreds of those we call 'their service providers'. These groups include government and private dairy specialists, who have all contributed in their own way to support their dairy farmer clients in their professional quests to increase their profits in order to improve their livelihoods as efficient converters of livestock feeds into a human food of the highest nutritive value and diversity of tastes. Thank you all for input into our most enjoyable travel and learning opportunities over the last 20 years.

We would also like to acknowledge the CSIRO Publishing team for the highly professional way they have put this book, John's sixth tropical dairy farming book, together and arranged for all his publications to be made freely available to all readers of tropical dairy production. This was all made possible through The Crawford Fund's generous financial support. John took many of the pictures during his recent trips to study the dairy industries of North Vietnam and Kenya, both in the tropics but with very different climates. The Kenyan smallholder farms were literally on the Equator but, being at high altitudes (1700 m), had constant comfortable temperatures year-round. In contrast, the Vietnamese feedlot farms although 18°N, and only 20 m above sea level, suffered severe winters and extremely hot summers. Two very different tropical environments and milk production systems that have resulted in vastly different blueprints.

We are also grateful to Kate Blasak, of the World Society for Protection of Animals, Sydney, and Phillip Schultz, of Daviesway, Kyabram for allowing us to use one of their photographs.

John Moran and Philip Chamberlain
August 2016

Acknowledgement of The Crawford Fund

The publication of this book would not have been possible without the generous assistance of The Crawford Fund whose mission is to increase Australia's engagement in international agricultural research, development and education for the benefit of developing countries and Australia.

THE CRAWFORD FUND

For a Food Secure World

Chemical warning

The registration and directions for use of chemicals can change over time. Before using a chemical or following any chemical recommendations, the user should ALWAYS check the uses prescribed on the label of the product to be used. If the product has not been recently produced, users should contact the place of purchase, or their local reseller, to check that the product and its uses are still registered. Users should note that the currently registered label should ALWAYS be used.

1

Introduction

This chapter presents an introduction to the material available in each chapter of the book as a guide to the reader on how to gain the most from the material presented.

1.1 Milk supplies and demands in tropical Asian countries

Globally, agriculture provides a livelihood for more people than any other industry (primary or secondary) while dairy farming is one of the major agricultural activities. The FAO has estimated world milk production in 2012 at 754 billion litres. Hemme and Otto (2010) estimated that 12 to 14% of the world's population (or a total of 750 to 900 million people) live on dairy farms or are within dairy farming households. Livestock provide over half the value of global agricultural output and one-third in developing countries. Milk is nature's most complete food and dairy farming represents one of the fastest returns for livestock keepers in the developing world.

The Asia–Pacific region has seen the world's highest growth in demand for milk and dairy products over the last 30 years. Even though Asia has increased its milk outputs (as a percentage of global production) from 15% in 1981 to 37% in 2011, it still accounts for over 40% of the world's total dairy imports. The consumption of milk and dairy products in Asia has doubled over the last 30 years, now contributing to more than 60% of the total increases in global consumption.

Many of these countries now have school milk programs to encourage young children to drink more and so improve their health through increased consumption of milk's energy, protein and minerals (particularly calcium and phosphorus). In future years, as these children grow and have their own families, milk consumption will increase at an even faster rate. According to Delgardo *et al.* (2003), per capita milk consumption in South East (SE) Asia is expected to nearly double from 10 to 12 kg/head/yr in 2003 reaching 19 to 20 kg/head/yr by the year 2020. This 3% per annum growth will lead to a total milk consumption of 12 million tonnes/yr by 2020, which Delgardo *et al.* (2003) predict will require 9 million tonnes of milk/yr net imports just to satisfy this demand. This is up from the 4.7 million tonnes of milk/yr imported in 2000. In summary, by 2020, SE Asia will then only be producing 25% of its total milk requirements.

Such growing demands have arisen by a combination of:

- increasing per capita incomes
- the emergence of affluent middle-class people in many low to middle income countries
- trends which increase the demand for protein foods and value-added dairy products
- increasing urbanisation
- expansion of modern retail outlets (with refrigeration cabinets) throughout Asia.

In other words, higher incomes and increasing urbanisation have combined with economic reforms and market liberalisation policies to heighten the milk import dependency of many countries in this region. Asia has then become increasingly dependent on the highly competitive, but ever increasingly volatile, global dairy commodity markets. Consequently, many Asian countries are striving towards self-sufficiency in dairy production, at least in drinking milk.

Most Asian countries still do and will rely heavily on imported dairy products for many years to come, even though many have active government policies to increase domestic milk production. There is a group of Asian countries with low per capita milk consumption and low self-sufficiencies and these countries are likely to be the ones with most proactive dairy development programs. These include Philippines, Indonesia, Thailand, Malaysia, Vietnam, Cambodia and Laos.

Dairy farmers can produce milk from six different types of ruminant, large ruminants (cattle and buffalo, plus camels in Africa and yaks in Asia) and small ruminants (goats and sheep). Small ruminants are rarely milked in Asia. Of the two buffalo ecotypes, river buffalo are the traditional dairy stock, with swamp buffalo rarely being milked. The majority of milk in Asia is then derived from cattle, with some buffalo milk produced in Myanmar, Vietnam, Philippines and Thailand while the large buffalo-milk-producing countries are India, Pakistan, China and Nepal.

While production has slowly responded to this growing demand, it has fallen short of consumption gains, leading to threefold increases in imports of milk and dairy products. Of extra relevance to this increase in demand, this region is home to two-thirds of the world's poor and undernourished people who will greatly increase their demands for dairy products in future years, for the above reasons. In the developing countries, over 80% of the annual global 240 billion litres of milk are supplied by smallholder dairy (SHD) farmers with only 1 to 5 milking cows.

Table 1.1 presents FAOSTAT (2013) data from 15 countries in South and East Asia on their per capita dairy consumption in 2009, their annual milk production in 2012 together with their changes in milk self-sufficiency over the last 10 years or so. With regard to changes in self-sufficiency of milk and dairy products, several countries have achieved close to 100% self-sufficiency, while others have been unable even to maintain previous levels of self-sufficiency because demand has greatly exceeded supply. Others have minimum levels that have hardly changed over the last 10 years.

Table 1.1. The size and self-sufficiency of selected Asian dairy industries in 2012.

	Total dairy consumption (kg/capita/yr) [2009 data]	Total milk production (million tonne) [2012 data]	Self-sufficiency in milk (%)			
			2000	2005	2009	2011
South Asia						
Bangladesh	20.2	3.5	74.7	66.7	70.5	87.5
India	72.2	124.8	100	100	100	100
Nepal	43.0	1.7	99.1	98.6	98.0	100
Pakistan	171.9	37.9	99.5	99.8	99.8	99.4
Sri Lanka	35.9	0.3	32.9	24.7	33.8	27.0
East Asia						
China	30.0	42.4	87.0	91.8	95.1	87.2
Japan	76.5	7.6	79.8	80.4	81.2	83.5
Mongolia	150.5	0.3	99.2	96.4	96.7	99.3
S. Korea	22.9	1.9	90.7	86.1	82.2	74.7
SE Asia						
Indonesia	11.5	1.4	37.6	31.9	30.0	39.6
Malaysia	36.7	0.1	3.3	3.5	6.1	6.8
Myanmar	29.0	1.6	83.0	85.6	92.2	94.4
Philippines	13.3	0.02	0.7	0.8	1.1	0.7
Thailand	21.9	1.1	28.8	45.5	50.3	51.4
Vietnam	11.5	0.4	19.5	19.6	30.6	32.6

1.2 An outline of this book

This manual covers a wide range of topics primarily related to achieving the sustainability of dairy production systems in tropical developing countries. This book is written for all the stakeholders in SHD production systems in the tropics (in Asia, Africa and Central America), but with an emphasis on the SE Asian countries (see Figures 1.1 and 1.2). In addition, unlike John Moran's previous books on tropical dairy farming, it provides guidelines for the best management practices (BMP) of large-scale, more intensive dairy production systems. While smallholders are the major suppliers of milk in the tropics, numerous larger farms are becoming established throughout the tropics to satisfy the increasing demands for fresh milk. It is hoped that both types of farmers and their advisers will gain much from this manual and improve the welfare, milk quality and productivity of their dairy herds. Policymakers and senior management should also benefit from reading selected chapters. In addition to photographs from Asia, this book includes several photographs from another tropical dairy industry, namely Kenya in East Africa.

Most tropical countries have proactive programs to increase local supplies of milk, which require increasing numbers of well-trained workers to service their dairy industries. Consequently, educators from agricultural schools, universities and technical colleges need to keep abreast of the latest technical developments and applications in dairy farming. This book also aims to serve this purpose.

The key target audiences for this book are:

- Farmers and stockpeople who want to improve the productive performance of their dairy herds.
- Farm advisers who can assist farmers to achieve this aim.
- Educators, usually at a technical level, who develop training programs for farmers.
- Educators, usually at a university level, who train dairy advisers in the basics of dairy production technology.
- Other stakeholders in tropical dairy production, such as local agribusiness, policymakers and research scientists.

These chapters are written so that they can be understood by advisers and tertiary students. As the trainers must ensure that other target audiences can comprehend their course material, they should select the most relevant sections to incorporate into basic programs for farmers. Each chapter has been written as a 'stand-alone' document, that can be individually downloaded from the internet. For this reason, there may be some degree of repetition between chapters but this has been kept to a minimum.

Every profession has its jargon, or words developed specifically for that profession, and agriculture is no exception. There are some very specific terms and acronyms that are routinely used by dairy researchers and advisers. These are explained in the Glossary and when they are first mentioned in this manual. Full publication details of all sources of information are presented in the References and further reading section, while the Appendices include several specific topics that

(a)

(b)

Figure 1.1: Two extreme systems for producing raw milk in the tropics. (a) Smallholder Kenyan dairy farmers delivering their morning milk to the local dairy cooperative collection truck near Nairobi. Each farmer averages 2 milking cows and sells ~5 L milk/farm/day. (b) Imported Australian Friesian cows resting in their dirt yard next to their free stall shed in North Vietnam. These cows average 20 to 24 L milk/cow/day.

Figure 1.2: Dairy farmers in Malaysia developing their individual farm blueprints during a Farm Business Management workshop.

can be easily accessed. For ease of finding specific information, the Index lists all the key topics covered in the book and their relevant page numbers.

It is nigh on impossible to write a book about tropical dairy farming in which every fact is relevant to every reader. What is of most importance to the actual farmer may not be the most crucial fact for the educator of technical or university level students or the dairy adviser who chooses to reads this book. Furthermore, tropical dairy smallholdings take many forms – ranging from very small farms with fewer than five milking cows that are all hand milked, to larger operations with say 30 milking cows that are milked using 'bucket milkers' – these could be all owned by the one farmer or constitute a colony farm with many farmers owning small herds. Although grazing the milking herd is a rare feature on most tropical farms, the larger ones with adequate land may be able to graze their cattle, especially their dry cows or yearling heifers. Herd dynamics in a grazing situation can be very different to those in the confines of a shed, particularly one based on tie stalls. In future years, farmers may expand and be able to incorporate a milking parlour with fixed in-line milking harvesting equipment. In fact most sustainable and profitable dairy farms grow in the medium to long term, so for today's farmer with 10 cows or fewer who may become tomorrow's farmer with 20 to 30 cows, their knowledge of and practices to optimise cow performance will need to be regularly updated. This book, although not aiming to be 'all things to all people' has been written for a very diverse audience, so there will be technical aspects that are less relevant to some but more relevant to other readers.

It is quite likely that many SHD (and even large-scale) farmers would find this book too technical and difficult to comprehend. If English is not their mother tongue, and unless they were well educated and/or travelled, very few may fully understand the subtleties of the English language. Future translations may become available in the future if there is sufficient demand.

1.2.1 What is a blueprint?

The title of this book, *Blueprints for Tropical Dairy Farming*, may make people wonder what it is all about. We are using the term 'blueprint' because much of this book is made up of checklists for many of the farm activities essential for the 'best management practices' (BMP) of tropical dairy farming. The senior author has previously written five books on tropical dairy farming (see the references at the end of the book) which detail much of the background and explanations of many of these BMPs. Rather than repeat the theory behind these BMPs, the reader is directed towards the relevant chapters in the previous books which are listed below.

1.2.2 The chapters of this book

The outlines of each chapter are as follows:

- Chapter 2 describes the diversity of dairy production systems in tropical Asia, and also provides a comparison of the pros and cons (that is, the arguments in favour of and against) of smallholder versus large-scale intensive dairy farming.
- Chapter 3 discusses the major benefits of intensification of smallholder dairy farming in the tropics followed by their key constraints with various approaches to their solutions.
- Chapters 4 and 5 present recommendations when considering the physical aspects (such as farm layout, shed design and farm infrastructure): first, of smallholder and second, of large-scale dairy farms.
- Chapter 6 discusses the selection and management of the human resources (farmer and farm staff) on both smallholder and large-scale dairy farms.
- Chapters 7 and 8 discuss the recommended BMPs: first, on smallholder and second, on large-scale, intensively managed dairy farms. Chapter 7 also introduces the concepts of Golden Rules and Key Performance Indicators (KPI) to provide the basis of a comprehensive capacity building program for smallholder farmers.
- Chapter 9 discusses the importance of farmers developing their observation skills when first entering the cowshed.

- Chapter 10 provides an insight into the problems associated with poor stock welfare, a potentially major issue with housed dairy stock in the tropics.
- Chapter 11 introduces the concepts of farm business management to provide farmers with a structured approach when assessing their business performance. This is followed with a case study of profits achievable on a large-scale intensive dairy farm in Indonesia.
- Chapter 12 discusses regional development programs, mainly from a smallholder dairy industry perspective. This leads to a complete regional development package called the Dairy Smallholder Innovation Program (DSIP), which incorporates five closely related strategies, several of which have been outlined in previous Chapters.
- Chapter 13 (the final chapter) highlights some of the developments still required for tropical dairy farming: first, *to become* and second, *to remain* more efficient and profitable. It also details how farmers can access through the internet a 16-module training program developed by FAO on feeding management of dairy cattle in tropical countries.

As many of the tropical Asian countries now have at least several large-scale dairy feedlots in their dairy supply chains, it would be remiss of us not to devote discussion to their BMPs, as well as successes and failures of such operations. In so doing, some topics in this book have been separated specifically into those relevant to smallholders and those relevant to large-scale dairy farms, for example Chapters 4 and 5 and Chapters 7 and 8.

Many of the chapters introduce new information, therefore it is important to include the background and other technical information to justify their recommendation as part of these blueprints. For example, during the 10 years since the publication of *Tropical dairy farming* (Moran 2005), there has been increasing emphasis on incorporating large-scale dairy farms into the domestic milk production systems of virtually every Asian country. Table 1.1 highlights the fact that in many Asian countries, dairy self-sufficiency has decreased (for example in Sri Lanka, Indonesia and South Korea) or has remained very low (in Philippines and Malaysia). Probably the two best 'success stories' regarding improving dairy self-sufficiency would be Thailand and Vietnam. In both these countries, large-scale dairy feedlots are proving successful in increasing local milk supplies.

1.2.3 Sourcing additional background and technical information on various blueprints in these chapters

In addition to new recommendations on improving the management of dairy farms, both large and small, this manual comprises many checklists based on background and technical information previously published in John Moran's previous books on tropical dairy farming.

The number of internet hits on John's books as of August 2016 were as follows:

A) 2005; *Tropical Dairy Farming: Feeding Management for Smallholder Dairy Farmers in the Humid Tropics* (68 301 hits).
B) 2009; *Business Management for Tropical Dairy Farmers* (124 936 hits).
C) 2012; *Managing High Grade Dairy Cows in the Tropics* (28 394 hits).
D) 2012; *Rearing Young Stock on Tropical Dairy Farms in Asia* (101 023 hits).
E) 2015; *Cow Talk: Understanding Dairy Cow Behaviour to Improve Their Welfare on Asian Farms* (21 850 hits).

All chapters of these books are freely downloadable from www.publish.csiro.au.

To assist readers in sourcing more background and technical information on the blueprints discussed in this book, the authors considered it beneficial to list the chapters in this book followed by the relevant chapters in the senior author's previous five tropical dairy farming manuals. As shown below, each chapter in this book is accompanied by a book code referring to the relevant chapter(s) in the previous books (e.g. A3 being *Tropical Dairy Farming*, Chapter 3).

1. Introduction; B3, E1.
2. The diversity of dairy farming systems in tropical Asia; A2.
3. Addressing the key constraints to tropical dairy farming; B2, E11.
4. Physical features on smallholder farms; A8, A20, C13, E7.
5. Physical features on large-scale farms; nil.
6. Managing the human resources of tropical dairy farming, nil.
7. Best management practices on smallholder dairy farms; C3, C18, D16, E5, E6.
8. Best management practices on large-scale dairy farms; A15, A16, A19, B14, C4, C6, C7, C8, D1.
9. Developing your observation skills; E5.
10. Addressing the problems of stock welfare; C11, E1, E2, E5, E9.
11. Introducing the concepts of farm business management; A17, B9, B11, C16, D15.
12. Regional smallholder dairy development programs; A20, B19, E11.
13. The final word; A20, B16, B19, D16, E9.

2

The diversity of dairy farming systems in tropical Asia

This chapter presents an overview of the different types of dairy farming systems and development strategies in tropical Asia, to sustainably increase milk supply.

2.1 Definitions of dairy farming systems

Dairy farming in Asia can be broadly classified into three major types of production systems (Moran 2017) as follows:

1. *Mixed farming*, in which milk production only contributes a relatively small proportion of total farm income. Many of these farms have evolved from essentially cropping enterprises to those where livestock production becomes more important. Milking herd sizes are generally quite small on these farms, ranging from fewer than five to no more than say, 20 cows.
2. *Essentially smallholder dairy farms,* where milk production has increased over recent years to become a major contributor to farm income. In many cases, construction of the dairy facilities and available land has evolved and may not be sufficient for future expansion. Milking herd sizes are very small, generally no more than five to 10 cows.
3. *Larger specialist dairy farms,* which were established primarily to produce raw milk. Dairy facilities on these farms have been better planned to satisfy the requirements for a predetermined number of milking cows. In most cases,

land would have been allocated to produce the required fodder for the planned herd size, although in certain cases agreements would have been made with surrounding farmers to provide the necessary forage base. Milking herd sizes on these farms would range from 20 to 100+ cows. Large-scale intensive dairy feedlots would also be included in this category that can have up to 1000+ milking cows.

The contribution of these various farming systems to the total milk produced in each country varies with population pressures and the demands for alternative land use, other than providing livestock fodder. However, Systems 1 and 2 contribute the bulk of the raw milk produced. The majority of dairy farmers are smallholders, with average herd sizes often as small as one to five milking cows. In fact, smallholder dairy (SHD) farmers produce over 80% of the developing world's milk, making a significant contribution to the annual world production. Despite their high profile in their dairy industry, there are relatively few large dairy feedlots in any one Asian country.

2.1.1 Supporting smallholder dairy farmers

National governments, international aid agencies or benevolent governments or agencies from developed countries have devoted, and are still devoting, a lot of resources to improving the productivity and profitability, hence sustainability of the SHD industries throughout tropical Asia. The focus is on sustainable intensification of SHD farming. The term 'intensification' requires clarification. In general terms, intensification is understood to be increases in efficiency for a unit of a given resource. For advisers and researchers of crop–livestock or pasture-based livestock production, the term is often interpreted as increasing productivity per unit of land, usually associated with an increase in stocking rate.

The national dairy development (5- or 10-year) programs in most Asian countries concentrate much of their efforts towards the System 2 farmers mentioned earlier i.e. smallholder dairy farms. In other words, they are trying to phase out 'part-time' dairy farmers (in System 1) and encourage 'full-time' dairy farmers. National dairy plans provide government support, which often includes the establishment of dairy cooperatives or even just installation of new milk cooling units in areas where dairy farming is being encouraged.

System 3 farmers (larger, specialist dairy farms) are usually less reliant on public support as their establishment is often financed by private investors. Furthermore, in many cases the farm and stock are not actually owned by individuals, but by a group of investors. However, in recent years there has been considerable interest (and investments) in larger-scale, feedlot dairies. This is occurring because governments have struggled to overcome the inefficiencies of current SHD systems, such as low milk yields, poor milk quality, poor cow fertility and high young stock mortality rates which drastically limit their ability to achieve self-sufficiency in dairy production.

Smallholder farms generally yield low outputs of milk per animal. However, on a cost–benefit basis, the use of by-products or other waste as feed, and multiple outputs such as calves and meat production, the continued efficiency of smallholder systems can outweigh the apparent efficiencies of dairying monocultures. Application of current technologies will allow increases in the production and efficiency of milk production by better understanding the nutrient requirements for milk production, in addition to those for growth and meat production.

As stated earlier, many resources from international aid agencies, benevolent governments or national governments have been allocated to improve the sustainability of the SHD industries throughout tropical Asia by improving the productivity and profitability of this sector. This support is ongoing. The success rate of such programs is very variable when assessing the achievements against their long-term objectives. However, SHD farming is still 'the backbone' of milk production throughout tropical Asia.

2.1.2 Types of dairy development

Simplistically, Staal *et al.* (2008) separated dairy development into two types, traditional smallholder and commercial large-scale industrial. These categories were developed for the convenience of global dairy policymakers and readers of this book would rightly argue that there are many 'in between' dairy development programs.

1. **Traditional smallholder systems** reflect small-scale farm household systems often associated with informal milk marketing systems that predominate in many developing countries. These are frequently based on mixed farming with cash crops and have small dairy herds with low levels of farm inputs and outputs. There are frequently nutrient deficits both on the farm and in the farm household. The farms are generally located close to the markets and consumers. Their milk marketing is diffuse with many small-scale market agents, based on labour intensive handling and transportation. The farm outputs are destined for mostly low return liquid products that are limited in diversity and with little emphasis on milk quality and food safety. In addition, there is great diversity in market behaviour and the farmers have limited input into dairy policymaking.
2. **Commercial large-scale industrial systems** represent large-scale 'industrialised' production systems and the integrated marketing observed in most developed countries. These systems are usually single enterprise with large herds and high levels of farm inputs and outputs. With high capital inputs, their production systems are based on economies of scale. They often have nutrient surpluses both on the farm and in the farm household. There is a high dependency on infrastructure (roads, water, electricity) and labour, as well as long market chains. Milk marketing is generally concentrated, consisting of few large-scale,

vertically integrated marketing agents and industrial processors based on capital intensive technologies with diverse products destined for value-added products, many non liquid. In addition, there is little diversity in market enterprise types with a larger input into domestic and international dairy policymaking.

As with all generalisations, there are invariably exceptions. Indonesia for example, is rather unusual in that its dairy production sector is based on traditional smallholder farmers but most of the milk is destined for industrial processing. Most of the farms are located in the highlands of Java whereas the milk is transported long distances to the coastal cities where the processing plants have been established, strategically located near the ports to more easily source imported milk powder.

Because it is much easier to 'showcase' large-scale dairies, they are popular with governments when planning high-profile regional dairy development schemes. However, they require a much higher level of well-managed technical input than do smallholders and once they become more heavily mechanised, 'more things can go wrong'. Hooten (2008) even goes so far to argue that in developing dairy industries, the growth and opportunity profile of such investments will eventually become stagnant compared to schemes more directed towards the poor, informal, smallholder dairy farmer.

2.1.3 Dairy herd sizes in Indonesia and India

Government statistics on dairy herd sizes generally only report the average number of milking cows on any one farm. The total number of dairy stock will be greater than this because farmers usually rear their milk-fed calves and weaned heifers on their farm. How many more animals constitute the total dairy herd varies but assumptions are that it is likely to be at least 50% more than the number of adult cows either being milked or non-lactating.

The following tables present data on the ranges in herd size in two of SE Asia's large dairy industries. Table 2.1 presents 2011 data for Indonesia (Morey 2011) while Table 2.2 presents 2015 data for India (Chamberlain *et al.* 2015).

Table 2.1. The number of farms with different herd sizes in Indonesia.

Herd size (cows per farm)	Average herd size	% farms	Number of farms	Number of cows
1–3	2.7	88	123 200	332 600
4–6	5.8	8	11 200	64 900
7–10	9	2	2800	25 200
10–20	17	1.5	2100	35 700
20–100	35	0.05	70	2450
100–1000	600	0.01	14	8400
Total	**–**	**100**	**139 384**	**469 250**

Table 2.2. The number of farms with different herd sizes in India and their annual milk yields.

Herd size (cows per farm)	% farms	Number of farms (million)	Number of cattle (million)	Average cow milk yield (L/year)	Milk production (mill MT)	% milk production
2	97	73	146	700	102	72
8	3	2	16	1100	18	12
18	0.4	0.3	5	1500	8	6
50	0.07	0.05	3	2500	6	4
100	0.03	0.05	2	4000	8	6
Total	100	75	172	2365	142	100

Nearly 90% of the farms in Indonesia have three or fewer milking cows (71% of the total cows) while 97% of the farms in India have only two cows (85% of the total cows). Furthermore, fewer than 1% of the farms in Indonesia and 0.5% of the farms in India have more than 18 to 20 cows.

2.2 Strategies to increase domestic milk production

National levels of milk production can be increased in various ways. These include:

1. Placing the highest priority on increasing the number of dairy farmers in the country, without greatly changing their average milk outputs per cow or per farm.
2. Placing more emphasis on increasing per cow milk yields in association with increasing the population of dairy farmers.
3. Increasing the number of cows (that is the average size of each milking herd) on any one farm, without greatly changing the number of dairy farms or yields of their milking cows.
4. Combinations of increasing the number of farms, the size of the milking herds and per cow production.
5. Changing the type of dairy farm, from the smallholder, but maybe part-time, dairy farms (say with one to five cows) to larger farms which are still privately owned (say with 20 to 100 cows) to a 'mega farm' that is owned by investors or other well-resourced individuals or commercial enterprises (say with 500 to 1000 cows).
6. Another approach is to reduce the wastage between the farm gate and the consumer's kitchen.

The first option requires sourcing new areas for dairy farm development, a task that is often difficult because of competition for existing land use and the current high density of dairy farms in dairying regions. One such example would be sourcing new land for dairy farming on the island of Java in Indonesia, where 97% of the dairy industry is already based. Developing new regions for dairy farming, for example outside Java, is a slow process as it requires:

- finding suitable highland areas and providing alternative locations for current landholders
- building milk-processing facilities near these new dairying regions
- sourcing adequate farm services such as water, electricity or gas
- sourcing suitable dairy animals to populate these new farms
- sourcing areas of fertile land to grow the required forages
- sourcing ample supplies of suitable by-products to provide the ingredients of concentrates for the dairy herd
- establishing a population of skilled workers who understand and can carry out the relatively sophisticated farm practices of successful smallholder dairy farming.

This last point is very important because increasing per cow daily milk yields from the current levels of say 8 to 10 kg/cow/day, to a potential 14 to 16 kg/cow/day (with existing genetic quality of most SHD milking herds) requires a set of skills rarely found in most SHD farming populations.

Mention of cow milk yields in Asia averaging 30 or more kg/cow/day in high genetic merit herds is simply unrealistic because of the many production constraints in the humid tropical climate (Moran 2013), even in the highland areas. Dairy cows only produce milk to their genetic potential when these constraints are essentially overcome. In Asia we have taken a temperate species of animal, the dairy cow, and expected it to be easily translocated to the foreign environments generally with high temperatures and humidities, often infertile leached soils and relatively constant day lengths.

Furthermore, low per cow milk yields are energetically highly inefficient because of low nutrient outputs produced each day in raw milk relative to the high nutrient input requirements for maintenance in milking cows (and growth of replacement stock), hence the inability to dilute these high energy requirements with copious yields of high energy raw milk. Therefore, depending mainly on increasing per cow milk yields to satisfy the high national demands for raw milk is a very slow and unreliable process.

The most obvious way to more rapidly increasing domestic milk supplies is through sourcing more dairy stock suited to the environment, assuming the land, feed supplies and skilled labour force are also available or can be developed. Natural increases of dairy cow populations cannot be relied upon to increase national cow numbers because of the high mortality rates of young stock and the poor reproductive performance of mature cows on most SHD farms in the tropics (Moran 2005). Some countries may develop national breeding centres to address these issues. However, the only reliable way to increase dairy cow populations in Asian countries is through importation of breeding heifers, either unjoined or in early pregnancy. This is certainly occurring in many countries in the region. These stock are available from developed dairy industries such as those in Australia,

New Zealand and the US and Europe. Some Asian countries, for example Thailand, also have stock available to import into other Asian countries.

These animals are usually exotic heifers, either unbred or up to 5 months pregnant, in that they originate from temperate dairy industries where they have been reared on pasture in a largely climatically comfortable environment. Upon arrival, they then have to adapt to all the constraints of tropical SHD farming, such as high temperatures and humidities, limited quantities of poor to moderate quality feed and the vastly different rearing environment of a low investment system with limited to no grazing and a small cowshed. Changes in animal behaviour clearly indicate that this adaptation period can be quite traumatic and lengthy, up to 6 months according to experienced SHD farmers. This is exacerbated by the often different standards of acceptable practices of stock welfare on their new home farm.

To find the most suitable animals, tropically adapted stock – or at least stock with some degree of *Bos indicus* (Zebu) breeding – are required or the temperate adapted heifers must be destined for highland regions where there are fewer constraints to profitable milk production. Even when imported into tropical highlands, they will be more susceptible to environmental and farm management constraints than the indigenous dairy stock. In addition, because the imported heifers will be invariably be of higher genetic merit than the local dairy stock, they require a higher quality of post-arrival feeding and herd management than would the local dairy heifers. Therefore, the higher level of skills necessary for satisfactorily managing these imported heifers means that the farmers, to which these animals are destined, need to be chosen and/or trained in such herd management skills.

2.2.1 The range in size of the milking herd on dairy farms

As already mentioned, most dairy farms in Asian countries only have relatively few milking cows. In addition to actual milking cows, the farms would also contain a few replacement dairy stock such as milk-fed calves or weaned heifers. Such farms often have other farming pursuits, hence dairy farming would not be a full-time activity for these farming families. The definition of a SHD farmer may include farms with milking herd sizes up to 10 or even 20 cows. In countries such as Indonesia, these farms may not have additional land on which to grow forages so the farmers would need to source the herd's forage requirements from grasses and other herbage from the roadsides, rice paddies or other areas (such as state forests or under cultivated cash crops) – generally freely available during the rainy season. Another source of forage for the 'landless' farmers would be from forage crops grown under the supervision of dairy cooperatives: in these cases they would not be free. Oil palm plantations are another source of forages recently used for grazing stock (generally beef cattle or young dairy heifers), say between 3 and 7 years after the palms are planted. If grazed any younger than 3 years, the young palm plants themselves are likely to be stripped of their leaves, while if the plantation is 7 years or older, there would be little sunlight reaching the ground, hence there would be negligible forage regrowth.

Other dairy farms range in herd size from 20 to 50 to 100 milking cows, and would still be privately owned. However, there is increasing interest in developing much larger dairy farms, containing 200 to 500 to over 1000 milking cows. Such farms could be considered as corporate farms, with ownership of land, facilities and stock by investors or other well-resourced individuals or commercial enterprises. These 'mega dairies' are constructed for, and generally achieve, increased per cow production of high quality milk. However, they are often criticised because their emphasis is more on the yields and economics of domestic milk supplies rather than the social and economic development of the country or region in which they are based. However, employing many farm workers on large farms will be of direct benefit to the local economy. In the long run, it is up to the particular Asian country to decide which is the higher priority, producing more and cheaper quality milk or facilitating the social development of the region.

2.3 The pros and cons of large-scale dairy farms

Large-scale corporate type dairy farming is not new in the developed world and is becoming more refined with increasing technical and commercial knowledge and experiences. Many of these current management practices can easily be transferred to the tropical, developing dairy industries, provided the local management is made more fully aware of the constraints to high levels of per cow milk production, cow comfort and fertility in such hot, humid environments with tropical animal health issues. These are certainly harder to manage in the tropics than in the temperate, developed dairying countries. Nevertheless, there is growing interest in establishing such 'mega dairy' farms in virtually every SE Asian country. Therefore, the design, construction and day-to-day management of such ventures justify closer investigation.

There are many 'pros and cons' associated with increasing milking herd sizes from 20 to 500 to 1000+ cows. A total of 15 pros and 19 cons, originally highlighted by Moran and Morey (2015), are presented below, together with a series of additional considerations.

2.3.1 Pros

1. Allows for mechanisation, hence reduced human error in everything from growing and sourcing forages to feeding, herd and effluent management to milk harvesting.
2. Provides enough cash flow for appointing experienced farm managers and other professional on-site staff.

3. Provides enough cash flow to justify routinely testing all feeds for their nutritive values.
4. Provides large volumes of shed effluent that can either be sold as fertiliser or used to supply many of the essential nutrients for soils to grow quality forages on-farm.
5. Potentially provides large numbers of bulls or steers for local dairy beef farming either by the dairy farm itself or by local smallholder beef farmers.
6. Supplies large volumes of high quality milk (or processed dairy product) which provide better bargaining and marketing opportunities.
7. Allows for more consistency with forage crop agronomy, hence more consistent forage quality.
8. Allows for bulk purchasing (hence cheaper) suitable by-products to provide the ingredients of concentrates for the dairy herd.
9. Requires investment in large numbers of livestock, which would have higher genetic merit and be more consistent than when sourcing local stock.
10. Mechanisation of milk harvesting reduces variation in milk quality.
11. Provides for opportunity for on-site milk processing to value-add the raw milk.
12. Provides raw products of better consistent quality more suitable for milk processing.
13. Provides employment opportunities for locals as farm workers.
14. Provides opportunity for local beef farmers or for dairy farmers to redirect their efforts from milking a small herd of dairy cows to become specialists in growing out the large number of replacement heifers required by the newly established farm. This may not eventuate because large-scale intensive dairy farms generally want to have direct lifetime control of their stock.
15. Provides opportunity to outsource other activities (such as contract growing of forages) and developing closer commercial and social relationships with the local farming population.

2.3.2 Cons

1. Requires access to large cash reserves to construct facilities and purchase farm equipment (such as for milk harvesting and preparing total mixed rations or TMR).
2. Requires access to large cash reserves to purchase livestock, generally in large numbers.
3. Requires specialist skills in land preparation, planting, maybe irrigating, harvesting and processing large quantities of forages.
4. Requires skills in design and construction of large sheds and other farm facilities.
5. Requires skills in ration formulation and other aspects of feeding management.
6. Requires skills in reproduction and other aspects of herd management.

7. Requires skills in addressing mastitis and lameness and other specific animal health issues.
8. Requires skills in animal health issues arising from heavy concentration of stock in one place.
9. Must improve local infrastructures (roads) to handle extra heavy traffic.
10. With poor effluent management, it could increase local pollution loads.
11. Needs access to large quantities of suitable by-products to provide the ingredients of concentrates for the dairy herd.
12. Ideally requires access to year-round supplies of quality forages.
13. May require additional skills and infrastructure for forage conservation, specifically silage making.
14. Requires skills in information and communications technology (ICT) for capturing data from many sources (stock, feed reserves, HR management) on the farm.
15. Requires ensuring all staff develop animal welfare friendly herd management practices.
16. With such large numbers of stock, any small mistake in farm management can have large-scale and expensive ramifications.
17. Biosecurity is of high priority for all equipment, livestock and staff.
18. Personality interactions between management, farm and administrative staff require close monitoring and if required, early intervention.
19. Integrity of management staff is paramount, to set best examples to farm and administrative staff.

2.3.3 Additional considerations

- There is likely to be a lack of firsthand experience in veterinary support for animal health, e.g. disease diagnosis, surgery (if required), recuperation and management. Animal health issues more specific to intensive management e.g. left displaced abomasum (LDA).
- Higher yielding cows have a greater susceptibility to heat stress, exposure to pathogens, poor feeding and herd management, hence they have a need for greater protection against such constraints.
- The importance of strict routines/protocols in feeding and herd management cannot be overemphasised, hence the need for more consistent support mechanisms such as maintenance and backup of forage choppers, feed mixers and feed-out machinery, as well as regular farm staff in service capacity building programs.
- The more rapid milk cooling and delivery of raw milk to processing plants will ensure better integrity of value-added dairy products.
- There is also a need to eventually develop public relations programs to counteract the propaganda disseminated by the animal liberation lobby and other anti 'industrial farming' social pressure groups.

- There is a complete lack of experience in the logistics of large-scale intensive dairy farming in contrast to more in the traditional dairy production systems.
- There is an obvious need to find investors who are prepared to wait several years before they can achieve positive cash flows and maybe even longer for eventual profits.
- There is a lack of consultants with experience in initiating and operating intensive dairy systems in the humid tropics, which is very different to temperate dairy feedlotting.
- There seems to be a lack of evolution of dairy genotypes adapted to the humid tropics yet can still achieve high feed intakes, hence levels of cow performance in less constrained environments.
- The most important objective of tropical dairy feedlotting is counteracting the deleterious effects of high temperature/humidity together with high internal heat production arising from high feed intakes (see Figure 2.1). Tunnel ventilation and fans/sprinkler systems are still ineffective at achieving thermo-neutrality.

Figure 2.1: No matter what the milk production system in the tropics, there is a need for additional ventilation to reduce the adverse effects of heat stress during summer. This picture is of a large free stall shed in North Vietnam with fans directing air into the free stalls as well along the feeding face of the shed.

- Year-round quality forage supplies normally arise through developing forage-cropping systems together with forage conservation (usually as silage).
- It is essential to install systems to cope with large amounts of shed effluent and its recycling to forage production areas.
- There is a need to invest and maintain reliable machinery to overcome constraints arising from variability in farm staff skills. Labour is cheap but often unreliable.
- There is a need to invest in equipment and materials and develop skills in sophisticated tasks associated with large-scale dairy systems, e.g. oestrous synchronisation to produce batch calving to overcome problems of poor conception rates in summer, embryo transplants to more rapidly increase herd genetic merit, routine feed dry matter (DM) testing to aid ration formulation, microscopes for assessing semen quality, Brix spectrophotometers to assess colostrum quality, feed additives with specific purposes (such as Betaine) to address heat stress.

2.3.4 Other aspects of large farm management

It is more desirable to gradually increase herd (hence farm) size over a period of years to allow for the staff to become more familiar with the principles and practices of large herd management. This will also provide better opportunities to observe and address any unforeseen key issues in such a large-scale intensive dairy enterprise.

Achieving the Key Performance Indicators (KPI) for cow performance (such as weight and age at first calving, milk yields at different stages of lactation, herd fertility, mastitis and lameness problems) and others listed in Section 7.4 of Chapter 7 are essential to achieve the required KPIs for farm and business performance (such as cost of production, gross profit, return on assets) within predetermined, realistic time frames.

As in all countries, there are likely to be government (national, provincial or local) regulations and incentives related to developing a large-scale dairy enterprise. It is then essential, to seek such information at an early stage of farm development.

It must be emphasised that large-scale dairy farming requires large amounts of readily available liquidity (or cash). Cash flows are likely to remain negative for several years. If sourcing dairy heifers to populate the farm, they will also take several years to reach maturity and attain adult levels of cow performance. For example, when developing a budget and projected cash flow on a farm with optimum feeding and herd management, it is best to gradually increase the predicted milk yields, even of high grade Friesians, over the first few years of farm operation from say 4000 to 5000 to 6000 to 7000 kg/cow/year consecutively for

(a)

(b)

Figure 2.2: Even on smallholder farms, there is a wide range in milk production per cow. These pictures were taken in Kenya on two free stall adjoining farms, highlighting the results of their variation of feeding management and cow comfort. (a) These Friesian cows were averaging 14.4 L milk/cow/day with each cow provided with a thick sponge rubber mat on which to lie. (b) These Friesian cows were only averaging 5 L milk/cow/day with no soft bedding. Note the difference in cow size and body condition between the two farms.

their first four lactations; this approach is discussed in Section 11.4.1 in Chapter 11. Furthermore, it is always better to err on the pessimistic side of projected cash flows until you are certain that all the anticipated (including some of the unforeseen) constraints to cow and farm performance have been identified and addressed. Intensive large-scale dairy farming is a long-term program so profits will arise in the near not the immediate future.

Dairy farming in the humid tropics is fraught with problems. However, once they are addressed, good profits can be made (see Figure 2.2). For example, Seruni *et al.* (2015) compared the financial performance of a large-scale (with 1467 head) operation and a small, well-managed farm (with 52 head) in the Chiang Mai highland region of northern Thailand, finding on both farms, positive profit levels. Although the capital costs were 65% higher per cow on the large farm, profits expressed as the income less costs as a percentage of costs, were 63% on the large farm versus 55% on the small farm. On the large-scale farm, daily income was US$7.49/cow, daily costs (both variable and fixed) were US$4.59/cow thus providing a daily profit of US$2.70/cow.

3

Addressing the key constraints to tropical dairy farming

This chapter presents an insight into the constraints that commonly limit tropical dairying productivity as well as the benefits and limitations of development and intensification of SHD tropical dairy farming.

3.1 Key activities on any dairy farm

On any dairy farm, no matter its size or location, the dairy production technology can be broken down into nine key activities, which can be considered as steps in the supply chain of profitable dairy farming (Moran 2009a). Just as any chain is only as strong as its weakest link, each step in this supply chain must be properly managed. Weakening any one link through poor decision-making can have severe ramifications on overall farm performance, and hence farm profits. In chronological order of their role in ensuring a profitable dairy enterprise, the 'links' are presented in Figure 3.1.

3.2 The benefits of intensification

There are many benefits in improved productivity and profitability of smallholder dairy (SHD) farmers. In addition to higher levels of milk production (hence gross returns) per cow and/or per farm, Falvey and Chantalakhana (1999) list the following:

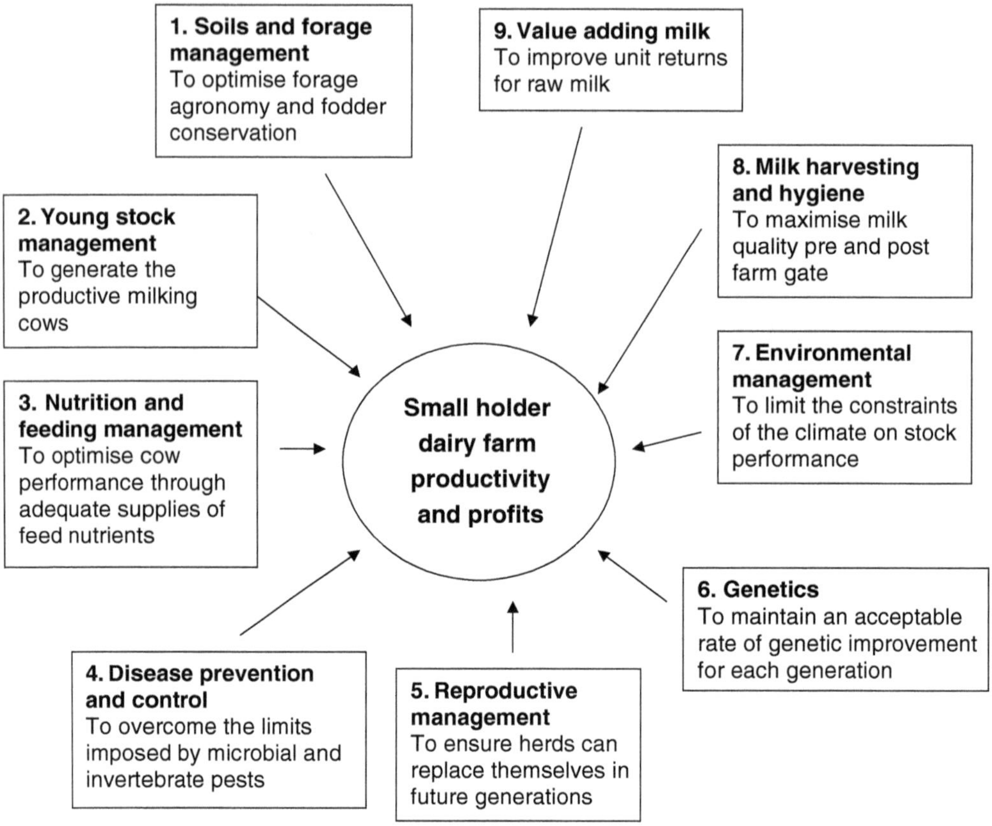

Figure 3.1: The nine steps in the supply chain of profitable dairy farming.

- year-round engagement of rural and peri-urban labour
- utilisation of agricultural and other by-products
- integration with cropping systems management
- conversion of by-products into organic manure for application to crops
- provision of nutritious and hygienic food for children
- production of meat from male calves and older cows
- reducing the cost of meat production for traditional markets as draught power declines as the primary bovine product
- a basis for rural and peri-rural industrial development through milk factories
- the development of new products for niche exports
- reducing rural to urban population drift
- draught and traction as a dairy industry by-product or adjunct
- landless people making a reasonable local living from dairying.

Prior to assessing the constraints to dairy production technology, it is worthwhile summarising a recent industry study of SHD farming in the tropics.

This study (Anon 2005) highlights the role of SHD farming, using a SWOT analysis to evaluate the industry's strengths and weaknesses. The analysis assesses the business or industry's strengths (S), weaknesses (W), opportunities (O) and threats (T). Although Table 3.1 refers specifically to Indonesia's SHD industry, it is applicable to any SHD industry in tropical Asia. Anon (2005) then concluded that SHD farming in Indonesia, as in other tropical Asian countries:

- improves the food security of milk-producing households
- creates employment opportunities throughout the entire dairy chain (for both producers and processors)
- is a powerful tool for reducing poverty and creating wealth in rural areas
- can incur relatively low production costs.

Table 3.1. Findings of a SWOT analysis of Indonesia's SHD industry (Anon 2005).

Components of SWOT	Findings
Strengths	Low production costs High farm income margins Low liabilities Relative resilience to rising feed prices SHD farmers are then cost competitive and resilient to market fluctuations They thus provide a competitive source of milk supply to imported dairy products
Weaknesses	Lack of knowledge and technical skills Poor access to support services Low capital reserves and limited access to credit Low labour productivity (small herd sizes and low output per cow) Poor milk quality and safety SHD farmers are often unable to take advantage of existing market opportunities
Opportunities	Growing demand for dairy products in developing countries Likelihood of increased milk returns Major potential to increase labour productivity Great potential to increase milk yields Employment generation Significant opportunities to improve the demand (quality and milk price) Significant opportunities to improve the supply (improving production technology) Empowerment of women Development of farmer and village cooperatives
Threats	Policy support in developed countries Exposure from competitive business forces Underinvestment in dairy chain infrastructure Unsuitable dairy development plans Environmental concerns such as a high carbon footprint Increasing consumer demand for food safety Succession of dairy farms Increasing local wage SHD rarely meets its full potential because of many threats, particularly the last four

However, in spite of several decades of dairy farming in developing countries, the productivity of SHD farms has remained relatively low due to a lack of appropriate dairy research and extension. Due to their socio-economic and agro-economic conditions being greatly different to those in developed countries, small farmers cannot readily adopt the science and technology available internationally. It is essential that any production technology being transferred is relevant to the needs of these smallholders as well as being feasible, given their local support networks of dairy cooperatives, advisers (government and agribusiness), creditors and milk-handling and processing infrastructures. Even the most appropriate technology is rarely transferred successfully to smallholders due to a lack of effective support services. There must be institutional support to facilitate dairy industry growth through mechanisms such as providers of farmer credit, farmer-training centres, well-equipped milk collection centres, processing and marketing facilities, farmer cooperatives or groups and appropriate research and extension infrastructures and methodologies.

For intensification to be sustainable, there must then be:

- adequate infrastructure and marketing opportunities
- access to reliable markets for increased milk production
- promotion of dairy development through government policy
- availability of credit for purchasing of livestock and planting pastures
- available productive and adapted forage species
- a supply of affordable cattle of the appropriate type
- ready access to information
- a supply of reliable and affordable labour
- farm management systems that ensure adequate feed throughout the year
- management of animal wastes
- health management and disease control measures
- adequate hygiene and management for milk collection.

3.3 Key constraints facing smallholder dairy farmers in tropical Asia

As a result of applied dairy research, development and extension over the last 20 years, Western countries have devised sophisticated dairy production systems (such as those described by Little 2012). Herd sizes have grown, efficient feeding, housing and herd management systems have evolved and many farmers routinely monitor test results on their cows for milk production, composition, quality and for mastitis. They then use this information for making decisions on nutrition, management and culling of milking cows and for breeding genetically improved stock. High labour costs have led to much mechanisation, such as machine milking and forage conservation, while grazing cows in SHDs can harvest their own forages far more efficiently than can farmers. Low population pressures, hence relatively cheap land, have allowed these farms to expand in both size and cow numbers.

Unfortunately the dairy industries of tropical Asia have failed to keep pace with the speed of such dairy development in Western countries (Devendra 2001). Cow numbers have increased in most Asian countries, largely through government support for social welfare and rural development programs. The increased demand for milk (accentuated through school milk programs) and the concept of national food security are the driving forces behind many dairy development initiatives. However, in terms of milk production per cow and feed inputs per kg of milk produced, improvements have been slow (Moran 2005, 2009a, 2012a).

Many of these developing dairy industries are located in tropical regions where high temperatures and humidity and, in some cases, seasonal growing conditions, adversely affect cow comfort and potential milk yields. Milking cows are not well suited to the tropics because their large requirements for feed nutrients, and their high internal heat production (compared to other species of livestock), cannot easily be incorporated into production systems that have to cope with poor forage quality, exposure to many disease agents and the climatic stresses that constrain cow comfort, immunity, appetite, reproductive efficiency, performance of young stock and animal health (Moran 2005).

In addition, many of the farmers, usually smallholders with fewer than 10 milking cows, have not been able to develop the skills of, or resources for, efficient milk production. As previously mentioned, this has primarily been due to poor extension services rather than lack of technical knowledge on tropical dairy farming. SHD farmers, with socio-economic and agro-economic conditions vastly different to those in Western dairy industries, cannot readily adopt the science and technology available in developed countries. It is essential that any production technology being transferred is relevant to the needs of smallholders as well as being feasible, given their local support networks of dairy cooperatives, advisers (government and agribusiness), creditors and milk handling and processing infrastructures (Devendra 2001).

Falvey and Chantalakhana (1999) categorised the factors limiting SHD production into:

- institutional factors, such as dairy cooperatives, suppliers of credit, training, extension services
- government policies, such as development programs, milk promotion, dairy boards
- socio-economic factors, such as farmer education, off-farm jobs, traditional beliefs
- technical factors, which can be further categorised into feeding, breeding, health
- post-farm gate factors, such as milk processing, marketing and consumption.

This analysis can be compared with a more recent study (Burrell and Moran 2004). In the early 2000s a series of strategic planning workshops was conducted in Indonesia to identify and prioritise the key constraints for milk production and to develop action plans.

Burrell and Moran (2004) summarised the priority industry issues in East Java and some of the action plans for industry development as follows:

1. *Low cow productivity*: improve management of feeding, reproductive management, cow comfort and milk harvesting.
2. *Low milk price*: reduce costs of production, improve milk quality, mediate on milk pricing, find alternative markets.
3. *Poor milk quality*: improve milking hygiene at both farm and post-farm gate, improve milk composition through better feeding management.
4. *Poor feed quality and availability*: identify better forage species (e.g. legumes), appoint quality control teams for concentrate supplies, utilise marginal land for forages.
5. *Cooperative management*: reduce management structure and merge small cooperatives, improve post-harvest technology, improve calf and heifer-rearing practices.

There were other industry issues raised but not discussed in detail. These included the need to promote fresh and manufactured dairy products, improve technology transfer, identify the positives to stimulate farmer motivation, work towards autonomy of cooperatives and improve collaboration between government agencies and training organisations.

A similar list of priority industry issues was developed independently for West Java, with some of the action plans for industry development summarised as follows:

1. *Human resources:* improve knowledge, skills and attitudes of farmers and support staff.
2. *Poor feed quality and availability:* increase area of land for growing forages, overcome seasonality of forage supplies, reduce variability of concentrate quality.
3. *Low capital investments in industry:* invest in infrastructure for post-farm gate industry support.
4. *Small scale of farming:* increase herd sizes, overcome shortage of breeding stock.
5. *Insufficient technology:* increase supply of breeding bulls, improve feed supplies, diversify farming systems, value-add milk in farming areas to help overcome farmers' low cash flows.
6. *Institutions:* improve coordination among service providers, introduce better control over milk quality, improve efficiency of administration in institutions.

Again other industry issues were raised but not discussed in detail. These included the need to promote fresh drinking milk, facilitate and support milk marketing and develop post-farm gate technology in milk processing.

The on-farm constraints to SHD production technology in tropical Asia are many and varied. Thirty-five key constraints were summarised by Moran (2013) and are listed in Table 3.2. They are categorised using the nine key activities from Figure 3.1 (see above) and complemented with a range of possible solutions to overcome each one. An extra category 'Other on-farm constraints' is included in this table to take into account those covering farm business skills.

Figures 3.2 to 3.4 highlight the contrasts in young stock management found in the tropics.

After several decades of dairy development in many Asian countries, typical milk yields per cow per day still range between 8 to 10 kg as compared to average yields of 20 to 30 kg in developed countries. In addition, the average calving interval of dairy cows on SHD farms is commonly as long as 16 to 20 months, when it could be reduced to 14 to 15 months. This clearly shows their low levels of farm productivity. Many technical solutions are available (as in Table 3.2) but they must be carefully selected so they will be suitable for small farmers and their socio-economic conditions. This means that scientists and extension workers must be able to understand all the factors influencing the acceptance when transferring such technology to farmers. Scientific knowledge alone cannot solve small-scale farm problems.

Socio-economic factors, including family and village structure, gender roles, family and individual goals, decision-making processes, financial drivers and drivers of change all must be understood and utilised in any new extension or farmer training process.

Policymakers should resist the all-too-common assumption that development efforts should move from smallholders towards supporting larger-scale, 'more efficient' milk producers to meet growing consumer demand. Instead, growing demand should be used as a stimulus to help continue and sustain SHD enterprises, particularly when they face increasing barriers to participate in value chain markets.

If well managed, SHD farms can compete with large-scale, capital intensive 'high tech' dairy farming systems as practised in both developed and developing countries. However, SHD development plans must include acceptable strategies to increase competitiveness in all segments of the dairy industry chain, namely input supply, milk production, processing, distribution and retailing. The future for SHD farming in tropical Asia is optimistic so long as the industry can rectify many of the constraints to improving domestic production of raw milk, particularly those at the farm level.

Table 3.2. Key constraints to improved milk production on tropical Asian smallholder dairy farms and possible approaches to solutions.

Key activity	Key constraints	Approaches to solutions
1. Soils and forage management	a. Low yields of forage	Use inorganic fertilisers as well as manure Reduce nitrogen volatilisation of shed effluent by directing it into water storage Optimise forage agronomy (soil preparation, weed control)
	b. Poor forage quality	Use inorganic fertilisers as well as manure Use most appropriate forage species for region Consider other forages such as tree legumes Reduce harvest intervals
	c. Shortage of dry season forages	Consider silage making of wet season forages Plan year-round forage supplies Utilise effective silage making and storage techniques
2. Young stock management	a. High calf mortality	Better parturition management to minimise likelihood of infecting newborn calf Ensure use of semen or bulls with low calf birthweights Improve colostrum feeding program (Quantity, Quality, Quickly) Pay greater attention to navel dipping with iodine Better shed hygiene Develop skills in identifying potentially sick calves Better health management Identify causes of death or sickness and change management accordingly Improve calf housing Minimise stress in calf shed Consider feeding less milk to encourage concentrate intakes Be more aware of fluid replacers v antibiotics for treating calf scours
	b. Poor post-weaning growth rates	Feed adequate amounts of concentrates Ensure calf concentrates have 18% protein Feed less forages to stimulate concentrate intakes Better health management Ensure routine Clostridial vaccination program Monitor post weaning growth rates Ensure free access to cool and clean water
	c. High wastage rates (from birth to conceiving in 2nd lactation)	Dairy cooperatives could consider heifer farms

3. Nutrition and feeding management	a. Low quality of by-products and formulated concentrates	Routine laboratory testing of ingredients and formulation Quality control during formulation Use coop system to bulk purchase quality by-products
	b. Poor performance of cows during early lactation (poor peak and daily milk yields, delayed cycling)	Ensure best forages for cows in early lactation, never rice straw Ensure enough forages are fed (30 to 50 kg fresh grass per cow per day) Monitor total dry matter intakes and increase if insufficient Consider wilting fresh forages to stimulate intake Ensure at least 16% protein in total ration Ensure all feeds are palatable Ensure free access to drinking water Provide Ca & P supplements in formulation Check if sufficient rumen buffers in concentrates Do not make concentrates into a slurry with water Chop forages to reduce selection and wastage Address any cow comfort and heat stress issues
	c. Cows (particularly high genetic merit cows) do not cycle for many weeks after calving	Ensure sufficient forages and concentrates are fed Check to see if rapid loss in weight or body condition Ensure at least 16% protein in total ration Consider vet checking for ovarian or uterine health Address any cow comfort and heat issues
	d. Seasonality of milk production	Plan year-round sourcing (growing or purchasing) of quality forages Ensure year-round supplies of by-products and formulated concentrates Ensure adequate supplies of drinking (and washing) water throughout the dry season Ensure adequate cow comfort throughout the year
	e. Little profits in milking cows	Check Milk Income less Feed Costs (MIFC) Be aware of marginal milk responses if feeding too much Set realistic target milk yields and feed to achieve them Ensure ration is balanced for nutrient contents Maybe feeding too many cows for available feed supplies Feed fewer cows better

(Continued)

Table 3.2. (Continued)

Key activity	Key constraints	Approaches to solutions
4. Disease prevention and management	a. Problems with lameness	Check floors for ease of walking on them Ensure appropriate animal husbandry and handling Consider footbath for all stock when required Check ration if too much concentrates causing laminitis Undertake locomotion test Identify and treat affected cows quickly
	b. Problems with mastitis	Ensure appropriate milking management Ensure milking machines (if used) are working effectively Identify subclinical cases with California Mastitis Test Ensure one towel one cow policy and towels are properly cleaned and dried Treat every infected cow with antibiotics ensuring withdrawal period is followed Milk infected cows last Initiate routine dry cow antibiotic therapy Consider culling chronically infected cows Ensure purchased cows are free of mastitis
	c. High calf and heifer morbidity and mortality	Follow procedures as in activity 2. Young stock management
	d. General animal health problems	Ensure appropriate animal health biosecurity plans are in place Develop skills in identifying potentially sick stock Routinely inspect stock for disease and external parasites Isolate sick stock Improve routine use of vaccinations Routinely use quality and viable pharmaceuticals Reduce the degree of exposure to infection by improving shed hygiene Consider testing for internal parasite egg counts Reduce any overuse of antibiotics Find well-trained and more practical veterinarians

5. Reproductive management	a. High age at first calving	Follow recommended procedures for improved post-weaning growth rates
	b. Low 100 day in-calf rate (pregnant within 100 days from calving) or high 200 not in-calf rate (not pregnant within 200 days of calving)	Better feeding management during early lactation Appropriate body condition score for stage of lactation Improve heat stress management and cow comfort Check AI techniques Can veterinarian confidently undertake pregnancy diagnosis? Pay closer attention to heat detection
	c. High number of services per conception	Improve AI techniques or check that technician is sufficiently skilled Pay closer attention to heat detection Consider vet checking for ovarian or uterine health
	d. Low % mature cows are milking	This is a simpler measure of poor reproductive performance so follow procedures above
	e. Increasing the proportion of heifer calves	There may well be a role for sexed semen in well-managed dairy farms
6. Genetics	a. Poor milking cow quality	This generally is not an issue because the genetic merit of imported dairy heifers is likely to be better than any cow on the farm, although more focus needs to be on tropical adaptability It is quite likely that the performance of most milking cows will be limited by environment (feeding, disease, heat stress etc.) rather than genetic merit Be aware of the genotype by environment interaction which means that high genetic merit stock require better levels of feeding and farm management to express their higher potential performance
	b. Most suitable genotype for the system	Some countries will not allow Jersey crossbreds or Brown Swiss to be imported, hence the imported Friesians limit the dairy production to the highlands If Jersey crossbreds and Brown Swiss are allowed to be imported, they may well prove the more profitable breed in lowland regions
	c. Difficulty of collecting robust data for genetic improvement programs	More emphasis on permanent identification of heifers Pay greater attention to maintaining cows in milking herds for relatively lengthy periods

(Continued)

Table 3.2. (Continued)

Key activity	Key constraints	Approaches to solutions
7. Environmental management	a. High incidence of heat stress during the 24 h period	Ensure shed construction is appropriate, such as roof height and access to breeze Count respiration rates to quantify degree of heat stress Pay closer attention to heat dissipation - Check shed design for ventilation - Consider artificial cooling (sprinklers and fans) Feed cows during the evening, when cooler Consider outside area for night-time cooling and heat (cycling) observations Feed better quality forages to reduce internal heat production
	b. High incidence of animal health problems due to poor bedding and shed hygiene	Improve shed hygiene Ensure bed is comfortable and not just cement Remove manure frequently Isolate sick stock
	c. Reduced forage quality due to high temperatures and rainfall	Unfortunately it is not easy since tropical forages are more fibrous than temperate forages. Soil testing can assist with overcoming monitoring leaching due to high rainfall. Harvest grass earlier
8. Milk harvesting management	a. Poor milk composition (fat and protein contents)	Address any limiting feed nutrient deficiencies and imbalances Ensure sufficient forage intake to maintain milk fat content Maximise cow comfort so cows will maintain their appetite and lie down for appropriate periods of time
	b. Poor milk quality (bacterial contamination)	Improve milking hygiene (hot water, detergent, sanitiser) Ensure machine milkers are operating effectively (short milking times, correct pulsation rate) Ensure rubber liners are correctly replaced Address any mastitis problems Ensure rapid milk cooling Could be a post farm gate issue hence outside farmer's control

9. Value-adding milk	a. Poor milk returns	Consider processing/value-adding to improve unit milk returns
10. Other on-farm constraints	a. The small farm size restricts development potential	Dairy cooperatives could develop cow colonies
	b. Poor profitability of dairy farming	Quantify profitability over 6 to 12 month period Quantify milk returns and overall farm income (actual and potential) Quantify Cost of Production (COP) Be aware that increased profitability can result from decreased COP as well as increased farm income Dilute fixed costs with higher farm cash throughput
	c. Low capital resources for investing in farm infrastructure	Seek alternative low interest loans
	d. Poor dairy farming skills	Institutional support to improve farmer training
	e. Underdeveloped entrepreneurial skills in dairy farmers	Work closely with potentially successful farmers to help develop these skills Provide training in farm business management and developing farmer business skills
	f. Poor farmer-management dairy coop relationships	Become more vocal to improve them

Figure 3.2: The constraint here is both genetic and feeding management. This dairy cow is still suckling her calf which is probably older than 9 months. Her Zebu ancestry will limit her milk producing potential while her calf would be drinking most of her saleable milk. The Kenyan farmer runs 13 beef stock plus two milking cows and only sells 3 L milk/day.

Figure 3.3: In comparison with Figure 3.2, the other extreme is this calf-rearing unit in a large-scale dairy feedlot in northern Vietnam. The Friesian calves are reared on calf milk replacer, as it is cheaper than whole milk, until 10 weeks of age with a target weaned live weight of 70 kg. Note the infrared heat lamps above the pens and the closed side walls, because winter temperatures and humidities can lead to cold stress in young stock. Clearly the environmental constraints have been minimised during winter.

Figure 3.4: Once weaned and fed a concentrate based ration until 11 months of age in small groups, the heifers are then run together as a large herd in the free stall shed and fed a high quality total mixed ration based on maize grain and imported lucerne hay. Any constraints in feeding management have been minimised.

4

Physical features on smallholder farms

This chapter provides an understanding of the important physical resources necessary for smallholder tropical dairy farms to be viable.

The following physical features are presented for a smallholder dairy (SHD) farmer milking say 10 cows, as well as rearing his or her own heifer replacements. If the ultimate plan for the farmer is to milk more (say 50 cows) or fewer (say five cows) stock, then the size of the physical features need to be adjusted in proportion to the size of their milking herd. This chapter deals with smallholder farms while Chapter 5 deals more specifically with larger-scale farms.

4.1 Land

4.1.1 Location

Factors to be considered when choosing land for dairy farming are:

- the presence of suitable fertile and uncontaminated soils
- no evidence of flooding during the wet season
- convenience for transportation within the area
- proximity to milk collection centres
- access to water supplies for farming and, of greatest importance

- year-round supplies of sufficient clean water
- highland areas because the air temperature decreases 6.5°C for every 1000 m increase in altitude above sea level.

Distance from factories (which may release bad odours or lead to pathogenic infections) and urban areas (where pollution from farm effluent will not be tolerated) should also be key influences on location. In addition, the availability of ground or surface water for irrigation and cleaning sheds, as well as for stock needs, all have to be considered when siting a dairy farm.

Ideally, the area of land allocated for the farm should be sufficient to grow all the forages required by the herd. This includes forages specifically grown for livestock feeding or forage by-products from other farm enterprises, for example from commercial crops, such as sweet corn or soybean. As it is generally in short supply, sufficient land for forage production should be available close by, or at least local farmers should be available and agreeable to grow the forages for sale. The greater the distance that forage sources are from the dairy herd, the more costly they can become and the more unreliable they are as regular farm feed inputs.

4.1.2. Forage production area

Area of forage grown

Moran (2005) developed a methodology to calculate the optimal area for forage production required to provide sufficient year-round forage supplies for a dairy farm. The calculated area required depends on key factors such as:

- The total yield of quality forages grown per ha per year, which depends on the agronomic skills in growing the forage crop, such as land preparation, weed control, using inorganic fertilisers in addition to cow manure and also harvesting the crop at optimum forage quality.
- The total amount of forages required over 12 months by each milking cow and replacement heifer.

The yields of forages grown are based on the typical yields of Napier grass under various levels of management, namely poor, average and good management are 67, 130 and 200 t respectively of fresh forage/ha/yr which corresponds to 10, 20 and 30 t DM/ha/yr respectively. Forage requirements for stock are 50 kg/cow/d fresh grass for milking cows (for 274 days in lactation), 30 kg/d for dry cows (for a 90 days dry period) and 20 kg/d for replacement heifers (for say 20 months).

For a farmer growing the maximum quantities of quality forages to feed his milking herd well, he should have no more than eight to 10 milking cows per hectare of forage grown on his farm. However, most farmers do not manage their forages sufficiently well to produce the highest yields of forage. Therefore, a more realistic recommendation would be six to eight milking cows (plus replacement heifers) per hectare of forage grown on farm.

This means that a 10-cow SHD farmer should aim to utilise a total of 0.75 to 1.0 ha for forage production. This recommendation is for a smallholder farmer in the humid tropics growing Napier grass as the primary forage source. This will obviously vary with the location and preferred forage species, but because of its relatively high yield, the required area is likely to be larger in regions where forage species other than Napier grass, are grown.

Types of forages that can be grown

Cook *et al.* (2005) developed a CD (entitled 'Tropical forages. An interactive selection tool') to select the most suitable forages that can be grown in different climatic regions in Asia, based on their optimum climate, soils, production system and management. These include both grasses and legumes. Growing tree legumes is another option. This requires additional planning in location of the trees, such as along roads or fence lines, and ease of access for harvesting.

Forages grown as single swards or as mixtures

Unlike temperate regions, it is not feasible to incorporate a forage mixture of grasses plus legumes into a single perennial forage sward. Different species of forages have different growing forms, or stem habits, namely: erect, prostrate, climbing, stoloniferous (stems growing along the ground) or rhizomatous (root nodules that convert atmospheric nitrogen into plant nitrogen). Erect forages will shade out prostrate forages, so it is best to grow each one as a separate crop. In addition, some forages are perennial while others are annual, requiring re-sowing each year. Others may be perennial but because of weed invasion and other constraints to production, they may require annual or biannual re-sowing.

Area available for grazing stock

Due to the shortages of suitable land, hence high land prices, Asian SHD farmers are forced into zero grazing production systems. Such stall feeding usually leads to less feed and water being offered, resulting in lower potential feed intakes than with grazing stock.

Stall-fed cows must be protected against the harmful effects of direct sunlight, mud, wind, and both external and internal parasites. However, such confined stock suffer more from certain diseases, such as mastitis, uterine infections and lameness, and also from poor heat detection, poor ventilation, dust, effluent and lack of proper bedding and poor cow comfort. Other benefits of zero grazing include protection against predators and thieves, reduced burden of day-to-day herding and lack of farmyard manure. However, labour requirements are greatly increased compared to grazing stock.

Allocating land for grazing dairy stock involves accepting lower yields of forage per ha because of the need for different grass species (than those used in cut and carry systems) and less efficient harvesting by the stock, compared to hand or machine harvesting. However, there are certainly benefits to stock wellbeing, and consequently performance, if provided with grazed forages. If land is available, grazing should be preferentially offered to growing heifers and dry cows, as they do not require twice daily attention as do milking cows and milk-fed calves.

Utilising shed effluent for forage production

Shed effluent can be used for fertilising forage crops in various ways. The most common is physically applying cow manure by hand. Nutrients from faeces and urine can be more efficiently recycled by applying all the effluent, diluted with water, directly onto the forage production area. Directing shed effluent into ponds reduces the loss of volatilised nitrogen from the urine thus enabling farmers to recycle more of this urinary nitrogen onto the forage crops. This system will require the construction of an effluent storage pond, and the installation of a recycling system such as a pump and hose, and/or a series of channels throughout the forage production area for distribution of the diluted effluent. If water is readily available for irrigating forages, such a distribution system could have a dual purpose. In dairying areas with available water for irrigation and suitable topography, consideration could be given to designing an irrigation system based on gravity (either flood or channels) rather than using pumps.

Other aspects of forage production and utilisation

Forages can be harvested by hand or machine (such as a mechanical brush cutter) to be chopped before feeding. It must then be transported to the cowshed, either by vehicle if distant, or by hand or hand-pushed trailer if close by. They may be wilted before being fed out to reduce moisture content that will stimulate voluntary forage intakes, in which case a series of wilting racks needs to be constructed close to the cowshed. Incorporating a plastic cover will provide protection against the rain to facilitate year-round wilting of freshly harvested forages. Finally, to reduce selection and wastage of grass stalks, they should be finely chopped, either by hand or machine, before feeding out.

4.1.3. Effluent storage

Shed effluent is quite bulky, as well as being difficult and unpleasant to collect, dispose of, or store and transport to the forage production area. Every day, each milking cow typically produces 30–35 kg faeces and 12–16 kg urine. In addition, most dairy sheds are washed with a hose, thus further increasing the volume of shed effluent. If rainfall is not directed away from any outside effluent storage, this run-off also contributes to the effluent load. A 10-cow (plus replacements) farm thus produces 0.2 to 0.3 t fresh manure and 120 to 180 L urine each day and this would amount to a total volume of 0.25 to 0.35 m^3/d, excluding any rainfall run-off.

On most of the smaller tropical dairy farms, the manure is collected with a shovel soon after it has been excreted and stockpiled near the exit to the shed while the urine and shed washings are directed away from this stockpile. Several times each week, this manure would be transferred into a wheelbarrow and taken out to the growing forages for distribution. As stated earlier, some farms direct all the faeces and urine into a storage pond for subsequent pumping out onto the growing forages. In some countries such as Pakistan, the shed manure is sun dried for use as kitchen fuel for cooking. It can also be used to produce biogas.

Most 10-cow farms would rely entirely on physical labour to collect and distribute the shed effluent. However, larger farms may be too big to depend entirely on physically carrying the effluent from the shed to the forage production area. Therefore, to minimise such labour, these farms would require a large storage pond plus manure pump and distribution system. The size of the storage pond would depend on the annual rainfall, the ability to divert run-off water to other disposal systems, the daily input from the dairy shed and the frequency of applying the effluent to the forage production area. However, it would have to have a capacity of at least 10 m^3 to provide for several days' storage.

The depth to groundwater under the soil surface is important as this will limit the pond depth and thus increase the required surface area. The base of the pond should be at least 1 m above the highest seasonal watertable. The permeability of the soil is also important, as this will influence any seepage into groundwater supplies.

The management of animal manure may represent a major sanitary and human/animal health hazard on SHD farms, with the problem increasing with herd size, unless specific facilities are constructed. The problem is associated with several issues:

- quantity and quality of faeces and urine produced
- adequacy and frequency of removal
- storage in proximity to shed
- labour availability
- methods of storage and disposal
- value and use of manure
- community concern about pollution (smell as well as contamination of groundwater and water courses).

The human health hazards are also becoming more serious than previously realised, due to inadequate supervisory and sanitary measures. For example, in Thailand a survey noted wastewater from SHD farms constituted a considerable risk to public health because of its very high contents of inorganic pollutants and the presence of coli-form organisms. Furthermore, there was considerable contamination of groundwater supplies in nearby towns via wastewater from local farms as well as via leaching from stored manure.

Not only is effluent management an environmental constraint to smallholder dairying because of its potentially adverse influence on cow performance and cost of forage production, but of equal importance, it can alter the livelihood of farmers themselves through community legislation limiting locations of farms. This can be particularly important for dairy farmers situated close to large towns and cities. In developed countries, there is now considerable effort to ensure farming activities are seen as 'clean and green' and this concept will no doubt extend into the humid tropics in years to come.

It is important to collect as much urine as possible from the shed. It is advisable to use minimum water to initially wash out the manure drains, then direct the main washings from the floor into another pit or direct out to the forage producing area. An ideal effluent system is to direct all manure and shed washings into the one pit equipped with a manure pump to direct the effluent to the forage producing area, along gutters between the rows of plants.

4.1.4. Water supplies

It is essential that dairy animals are provided with adequate supplies of good quality drinking water. If this is not readily available year-round, then a system should be developed for collecting rainwater during the rainy season and storing it until required. Throughout the humid tropics, greater attention needs to be given to ensuring all sheds and houses have gutters and downpipes to facilitate the collection and storage of clean drinking water.

Lactating dairy cows in the tropics require 60 to 70 L water each day for maintenance, plus an extra 4 to 5 L for each litre of milk produced. Water requirements rise with air temperature. An increase of 4°C will increase water requirements by 6 to 7 L/day per L milk produced. High yielding milking cows can drink 100 to 150 L water/day during the hot season. A 10-cow herd would then require up to 1500 L/d to ensure limited water intakes do not restrict appetite and consequently production. In areas such as Central Java, where the lengthy dry season leads to very deep underground bores being the only reliable source of stock water, the high cost of accessing this water often leads to shortages in drinking water, which can reduce feed intakes, accentuating the problems of reduced milk yields arising from low nutrient intakes. All too common, cow performance (milk yields and fertility) is limited by the shortage of clean drinking water.

Good quality water is also important for the effective use of the detergents and sanitisers for washing milk-harvesting equipment. This is essential in maintaining good milk quality. Non-potable water supplies must also be sufficient year-round for effective shed cleaning.

4.2 Buildings

It is essential to provide good housing and farm layouts for rearing and milking dairy stock in the tropics, for ease of cleaning out sheds and to reduce the extreme effects of air temperature and humidity. Such control improves milk production by reducing stress and disease hazards and also by making herd management easier. Environmental control includes sanitation and effluent management.

Good dairy sheds have many simple design features to improve cow and farmer comfort, hence cow performance and farm profitability. In hot climates, the following features are particularly important:

- sufficient well-drained yard space
- plentiful supply of water
- comfortable and dry resting area
- high roof, allowing breeze to blow through and reduce radiant heat
- sufficient water and feed trough space for each cow
- shade over the feeding as well as over the loafing area
- wire fences instead of solid walls so as not to restrict air flow
- planting trees to lower temperatures
- sowing grass or other forages in the area surrounding the shed
- white painted buildings
- shading over drinking troughs to keep drinking water as cool as possible.

It is often just as expensive in the long term to renovate old buildings that have been originally built for other livestock species or for a smaller dairy herd, than to remove the building entirely and start with a greenfield site. For example, converting an old pig shed into a dairy cowshed is likely to create problems in the future due to inappropriate shed height. A roof that is too low means the effect of solar radiation heating on the cattle is increased and it also results in ineffective ventilation. The foundations may still be adequate so could be reused, but the rest of the building is unlikely ever to be well suited to milking cows.

When designing buildings, consideration should be given to future expansion plans by ensuring there is sufficient space around them, as well as the adequate provision of utilities such as water, electricity and shed drainage. There may be regulations regarding buffer distances, that is space between buildings, which can restrict the ability to expand building layouts. Furthermore, effluent, odour and noise originating from cowsheds should have minimal impact on neighbours. The superstructure consisting of beam constructions, roofing and possibly walls, should be built preferably from local materials and follow local designs. There are several options for constructing the roof, such as iron, bricks or even bamboo/mud in a low rainfall area.

Soils should be evaluated to assess their suitability for constructing the buildings, and strong enough to carry the heavy loads of machinery, building foundations and, of course, the stock. During construction, better compaction can be achieved by removing and stockpiling the topsoil for later use for landscaping or adding to the forage production area. The proposed cowshed should be conveniently located close to the sources of home-grown forages.

Ideally buildings should be located above the 1 in 100-year flood level and all earthworks should be designed to avoid offsite impact of floodwater discharge either through funnelling or backwater effects. The natural drainage regime of the area must be considered so that uncontaminated rainfall run-off can be easily diverted away from the effluent storage. As already mentioned, gutters and downpipes could lead to drains or water storages to allow more effective control of such run-off.

The ideal aspect of the buildings depends on its geographical location. Orientation of buildings will impact on the severity of hot or cold environmental stresses at different seasons during the year. Buildings should be orientated to either maximise or minimise exposure to climatic elements such as prevailing winds, sunlight and rainfall. Locating the shed perpendicular to prevailing winds aids cross-ventilation. A north–south orientation promotes drying in exposed areas because the shade moves across the shed during the day.

4.2.1 Cowshed

Dairy facilities should be designed, constructed and maintained to minimise obstructions and hazards that have the potential to cause distress and injury to stock. Floors should provide satisfactory footing on non-slip surfaces, be well drained and easily kept free of mud and manure. Fences and gates should allow good animal flow and prevent injury. Crushes and head bails should allow efficient handling of stock without endangering them and allow their easy release. Sufficient space should be provided to prevent discomfort and ensure animals can meet their normal behaviour requirements for lying down and resting, moving around, eating, drinking and eliminating faeces and urine, and all without excessive social pressures.

Shed designs should ensure adequate air circulation, temperature control and removal of excess ammonia, carbon dioxide and slurry gases. Lighting should be adequate to enable inspection but not so intense as to cause discomfort. Animal groupings should be kept to reasonable sizes with stock sorted into different size or age groups to minimise social stresses such as bullying.

As all cowsheds need to have sloping floors to aid effluent removal, there should be sufficient soil depth to accommodate the cut, fill and borrowing requirements necessary for undertaking the earthworks. The shed should be formed well above the natural surface level to promote such drainage, as well as increase air movement. This will decrease shed odours and discourage insects. The cost of sourcing and moving sufficient suitable soil needs to be taken into account. Buildings should not be located on steep slopes, greater than 20% (1 in 5).

The surface should allow for effective drainage and prevent wastewater reaching the subsoil. Minimum gradients are 0.2% (1 in 500) for compacted earthen surfaces whereas concrete can drain at a slope of 0.5% (1 in 29) to 4% (1 in 13). From operating experience, slopes should be within the range of 2% (1 in 29) to 4% (1 in 13). Slopes can be steeper but these require a higher standard of construction and operational management. Excess run-off and sediment transport can occur, causing potential erosion on exposed areas as well as pollution problems. Specifications for concrete strength and thickness should be followed.

Shed design

The essence of good shed design is making the best use of natural ventilation, with open sides, a high and well-pitched roof together with an open vent along the top. The presence of cobwebs under the roof is indicative of poor ventilation.

The following discussion is for dairy farms located in tropical regions where the emphasis in shed design is to minimise heat stress. It is worth repeating that sheds should be constructed to maximise natural cross-ventilation, with open sides, a high and well-pitched roof together with an open vent along the top.

The shed should be sited so that breezes are not blocked by any obstacles or other buildings. Ideally, the shed should be on the highest ground possible, which will also be good for effluent drainage, with other buildings located downwind on the site. There should be a minimum of four times the height of the nearest wind barrier as a horizontal separation.

A north–south orientation is preferable to allow the sun to dry underneath both sides of the shed. An east–west orientation is not recommended, as the southern side will have less opportunity to dry out. Trees should be planted and protected from stock damage, on the western side of the shed to reduce solar radiation. Shadecloth that blocks 80% of the light can also provide protection as long as it does not interfere with ventilation within the shed. Eaves that extend one-third the side height also provide good sun protection.

White painted buildings reflect the solar radiation better than dark painted buildings. Reflecting roof materials such as galvanised iron or aluminium are good long-term investments. Insulation under the roof or solar panels on the roof, can reduce the heat load. Spraying water on the roof may only be effective in reducing roof and shed temperatures in areas with low humidity.

Effluent disposal systems

The floor of the cowshed should be made from concrete or easily washable material and designed for efficient drainage with a good slope with wide channels for easy urine and faeces removal. As previously mentioned, an effluent pond should be dug large enough to hold the shed's manure and shed wash for at least two or three days without emptying. A channel should be dug leading from the walking area to the pond and ideally lined with concrete. If manure is stored in a pit, it should be covered with a plastic sheet or banana leaves to reduce the sunlight, as this volatilises the nitrogen in the manure, reducing its value as a fertiliser. This will also allow the installation of a biogas plant during its construction or at a later date. A fence around the pit will minimise risks to children and wandering stock.

4.3 Farm infrastructure

4.3.1 Cow stalls

There are two basic types of housing, tie-up housing (tie stalls) and loose housing.

- With *tie stalls*, each cow is restrained in a stall. Feed is delivered in a trough in front of the cows. Milking usually takes place individually in the stall, by bucket or machine (bucket or pipeline milking). Manure is collected in a gutter behind the stock. As cows tied up all year round can suffer from feet problems and become stiff, they should be provided with an easy-to-clean, soft surface on which to lie down, such as rubber mats or mattresses. Providing cement only to lie down on will lead to excessive periods of standing and reduced appetites, which can eventually result in acidosis and lameness, hence poor milk yields and reproduction. Heat detection demands more attention in tie stalls while a high incidence of trampled teats can become another problem. Throughout Asia, dairy cows have been traditionally housed in tie stall sheds.
- With *loose housing*, cows are not tied up and can walk around freely. Such systems usually have a loafing area and require a comfortable and dry lying area, with the feeding area separated from the lying area. As the cows are forced to walk frequently, the manure is spread over a large floor area so has to be collected by scraping the dung by hand (or sometimes mechanically) into a manure pit. With adequate water supplies, rapid flushing of large amounts of water can clean alleys, directing effluent into a pond. With larger herd sizes, milking is often carried out in a separate milking parlour. The feed trough is separated from the loafing alley by either a feeding rack or wire rope.

Loose housing can be of two types, either with a common lying area with open lounging or with cubicles or free stalls.

- In *open lounging* systems, cows can lie down anywhere. The floor can be earthen or cement, generally with bedding material, the base being well drained. In dry climates, earthen floors with or without bedding can be used as long as the floor is turned over daily. This is called a 'compost barn' where the manure is sometimes cleaned out every day and the ground is turned and becomes hot, thus killing all the pathogenic bacteria. These areas must be

managed well to ensure they do not become too wet which can result in increasing mastitis loads. The area directly behind the feed troughs should be cement and at least 3 m wide. Each cow should be allocated a total of at least 9 m^2 resting area.
- With *free stalls*, each cow is provided with a cubicle, which she may enter and leave at will. Cubicles can be arranged in a single row or in more than one row with a central feeding alley or with feeding alleys along the sidewalls. The cubicles can be arranged with cows facing one another (head-to-head) or the other way around (tail-to-tail). With the tail-to-tail arrangement, a central loafing alley with a width 2.2 m between the cubicles is needed. If the cubicles are head-to-head, two loafing areas behind each row are necessary. Usually one of the loafing alleys is combined with the feeding alley behind the feed troughs.

Free stall layouts

The size of the cubicles depends on cow size. For cows:

- weighing 400–500 kg, they should be 104–109 cm wide and 198–208 cm long
- weighing 500–590 kg, they should be 109–114 cm wide and 208–218 cm long
- weighing 590–680 kg, they should be 114–122 cm wide and 229–244 cm long.

The surface of the free stall should have a consistent fall of 4%. It should:

- be comfortable to the cows to encourage high occupancy
- prevent hock damage and other injuries
- be easy to clean and be durable
- be cost effective to install.

The floor of the cubicles can be earthen with a thick layer of bedding, or can be solid, with a soft top layer. The simplest bedding is packed earth or sand, this often being inexpensive but needing daily care to maintain a flat surface. Sand is quickly pushed around by cows, so ideally, it should not be used with mechanical or liquid manure handling systems because it fills up storage tanks and is very abrasive, damaging equipment such as manure pumps. However, a sand trap can be installed, to allow the heavier sand particles to settle out before handling the other shed effluent.

A concrete foundation with a disposable bedding of chopped straw (and sometimes with the inclusion of lime and water), sawdust, wood shavings or crushed corncobs is more common in Europe, as rice hulls are not readily available. Rice hulls make ideal bedding for free stalls but their high silica content can also damage liquid manure handling equipment.

To prevent the cow from soiling the cubicle, shoulder, breast and head rails are needed. This provision will force cows backwards when they stand up. The distance of the adjustable shoulder rail to the back of the cubicle, measured diagonally, should be ~1.8 m, the height to the cubicle floor may vary between 90 cm and 1.05 m. It should be stressed that stalls must be designed to suit the average cow size and up-to-date information must be sought on stall design before any construction begins.

Open lounge housing create its own problems of regularly removing and cleaning the bedding and ensuring all cows will use it in preference to lying on dirty cement walkways, and hence increasing potential mastitis problems. Bedding must not be allowed to become wet from leaking water troughs or rainfall. Compost bedding systems vary greatly in their effectiveness and require constant monitoring. Some require daily input of straw and twice daily cultivation to maintain the correct moisture content. Others on sandy soil can be managed well by daily manure removal and occasional input of straw and daily cultivation.

Concrete alleys of loose houses should slope in the direction of the manure movements with a 3% slope and level from side to side. They must be easy to clean and allow for quick and free drainage with all effluent directed towards the storage pond, either for recycling as fertiliser or for initially producing biogas. In countries where dried manure is a value-added product for kitchen fuel, floor designs may differ slightly to those in other countries, where all effluent is directed to open drains. Effluent can also be directed towards a closed tank to be used for biogas production.

4.3.2 Problems with tie stalls

As far as the cow is concerned, tie stalls often:

- are extremely uncomfortable and have a hard surface
- cause pain, with risk of inflammation of knees and hocks
- are frustrating, because lying down with the head constrained is more difficult and the animal cannot lie down with a stretched foreleg or with its head tucked into its front legs
- are boring, because they restrict any opportunity for social contact or interaction and foraging
- restrict the ability for each cow to self-groom, particularly if the tether rope or chain is too short
- reduce lying and resting time, which can lead to increased lameness and stress
- can lead to abnormal behaviour, such as swinging from side to side and shifting/moving back and forth. Many cows often stand in a dull state if not eating or otherwise stimulated
- can lead to more frequent occurrences of stereotypical activities, such as bar biting or tongue rolling, which disappear when animals are transferred to loose housing

- remove the ability for stock to find the best location in the shed for reducing climatic stresses
- reduce the quality of the microclimate (higher relative humidity and air ammonia concentration and lower air movement), which can lead to reduced cow comfort and performance
- are all too commonly associated with poor hygiene, meaning that the cow is forced to lie in her own manure, thus creating additional problems with animal health and milking hygiene, such as mastitis
- can adversely affect reproductive performance, partly because cows stand for long periods, cannot move freely or assist in identifying other cows on heat
- can lead to higher teat injuries due to physical damage
- can have dimensions that may influence cow behaviour, cleanliness and wellbeing.

If circumstances require that the animal is temporarily tethered, she should be tied with a halter and released as soon as possible. Nasal tying, that is threading a rope through a hole in the nasal septum, can cause irritation and infection. A temporarily tied or single housed animal must always be able to keep at least visual contact with her herd mates to lessen stress reactions. A mirror may even substitute for other cows, for a short time. Some behavioural scientists argue that cows in tie stalls are considered to have had regular exercise if they are released for just 1 hr/d for 2 days per week. But one could likewise argue that 'the more exercise the better'.

If farmers persist with tie stalls, they need to pay closer attention to stall dimensions and bedding, because

- shorter stalls are associated with more cows with dirty hind limbs and rotated hind claws
- low tie rails can lead to more cows with neck lesions and broken tails, the latter resulting either from the tail being stepped on or forceful manipulation by the farmer
- short tie ropes or chains can lead to more swollen hocks and dirty hind limbs with cows struggling to stand and lie down easily
- hard bedding will lead to long standing times, acidosis and lameness.

In summary, tie stalls are *not* appropriate for the optimum wellbeing of milking cows and these conditions can negatively affect milk yields, reproduction and cow longevity. Tie stalls have been the traditional method of maintaining dairy cows in some European countries for many decades. This has created so much public concern that certain countries, such as Norway, have recently legislated against their future use in preference to loose housing.

4.3.3 Other facilities in cowsheds

Considerations in flooring

The required floor slope depends on the cleaning method adopted and the natural slope of the site. A minimum slope of 0.5% (1 in 200) is required to prevent pooling of water. Steep cow alleys (greater than 1 in 15) will result in cows not standing perpendicular to the feed trough, but standing uphill. Flood washing of the cow alley ideally requires a slope of between 2.5% (1 in 40) and 3.0% (1 in 30), to achieve acceptable wash with minimum amounts of water. If the slope is less than 2.5%, the volume of flood wash water has to be increased to get extra water depth to compensate for the lack of slope. There should not be any sideways slope on cow alleys with flood washing. If flood washing is not used, the cow alley should have a 2.5% slope away from the feed trough. Hose washing and dry scraping are not slope sensitive, so the site slope will influence the alley slopes. The feed alley needs a concave floor (with 50 mm fall) sloping away from the feed to the centre of the alley to prevent run-off water mixing with the feed. The lengthwise slope will be the same as the cow alleys.

Cows can be standing or walking on shed floors for up to 10 hr/d (Moran and Doyle 2015). In tie stalls, cows should be provided with rubber mats to cushion the hard concrete floors. With loose housing, concrete floors should be roughened to reduce slippage but not too rough so that they wear out hooves and lead to lameness problems. Another alternative is to install grooves or patterns in new concrete floors. These can even be cut into old concrete floors.

Grooves should be positioned so that at least one of the four feet lands in a groove whenever cows put their feet down to stop slippage. Parallel grooves should be spaces ~15 cm apart, 12 mm wide and 12 mm deep. Although placing grooves perpendicular to the length of the alley will maximise their effectiveness, they may compromise removal of shed effluent. So they should be located parallel to normal alleyway cow flow.

Anti-slip aggregates can be mixed in with the top layer of concrete or epoxy floor coatings can be applied to existing floors. Another alternative in areas of high cow usage, such as in feeding alleys, is constructed in-place rubber floors, such as those used on athletic tracks.

High traffic areas of concrete, especially in areas where cattle turn, should have rubber matting to reduce hoof wear.

Water troughs

Stock must be provided with sufficient drinking water at all times and the system should be able to supply at least 20 L/cow/hr to meet likely peak demand. Peak demand often coincides with feeding times and just after milking. Cows often drink a third of their daily water intake in the hour after morning milking on a hot day. The water supply pipe diameter needs to be at least 75 mm with an operating head of at least 10 m. The optimum temperature for drinking water is 15–17°C. Water troughs should be well separated (but within 15 m) from feed troughs with the same water trough servicing adjacent pens. Trough design should allow for regular, easy cleaning with a removable bung for complete drainage. Watering points should be cleaned out at least weekly to remove any feed and other contaminants. Water troughs should not be sited over or near bedding areas in sheds.

Each cow should be provided with 75 mm of linear watering space in free stall sheds, while for circular water tanks, one watering space (60 cm of tank perimeter) should be available for every 15 to 20 cows. A water depth of 15 to 20 cm helps keep water cooler, fresh and easier to clean because less debris accumulates. The optimal trough height is 60–90 cm, from ground to the top of the trough.

With loose housing, water troughs should be well separated and close to the concrete feeding area with provision for water spillages, or leakages, to directly enter the drainage system with no spillage onto the compost bedding area. Troughs should be located at the high point of a water line to reduce sediment and to facilitate purging air from the pipeline. In free stall sheds, they should be located at crossovers to reduce the incidence of stock blocking each other in alleys.

If cows are tethered, it is possible to install individual cattle drinkers, with two animals sharing one drinker. To ensure water is freely provided, troughs need to have float valves rather than depend on a hose whenever the farmer decides to refill it. A single watering trough system can control water levels in up to 10 small troughs just by one float valve in a central reservoir with syphon pipes leading to all the troughs.

Upgrading the facilities can involve increasing the rate at which troughs fill or providing greater trough capacity. Options involve upgrading pumps and water pipe diameters or increasing the number of watering points, particularly those close to the feeding area.

The tradition of mixing water with the concentrate portion of the ration is widespread throughout Asia. Farmers wrongly believe that cows will better utilise concentrate slurries compared to dry mixes. This may be one way of providing additional water, in housing systems where there is limited free access and stock are only provided with restricted times to drink. However, given free access to water, there is no need to feed slurries of concentrates.

Feed troughs and feeding strips

It is important that all stock can eat comfortably with minimal competition. Whether feeding into troughs or onto cement feeding strips, 70 cm feeding space should be allocated per mature cow. For a feeding strip or trough when cows eat from both sides of the strip, hence face each other to eat, this equates to a total feeding strip allocation of 35 cm/cow. Trough space can be reduced to 45 cm/head for 6-month-old cattle, or to 55–60 cm/head for stock 18 months old.

With feed troughs, the width should be related to the reach of the animals which is up to 80 cm for mature dairy cows eating from the base of a trough 30 cm above ground level. The base of the trough should then be raised 10 to 30 cm, with the front 50 cm above this. Cows in tie stalls are likely to have restricted reach. In loose housing systems, lockable feeding head bales will reduce the incidence of cows throwing feed around as they lift their heads, this means less wastage.

Feeding strips should be positioned 7 to 15 cm above cow feet level, with a nib wall 40 to 55 cm above the feeding strip, to protect the feed from the effluent. Feed barrier walls wider than 15 cm can restrict how far stock can reach into troughs or onto feed strips.

Feed troughs and strips should be situated on the high side of the pen, running parallel with the contour to minimise pad drainage. They should also have smooth surfaces, as those without grooves or holes that can trap feed are easier to clean and help reduce build-up of waste feed, mould growth and unpleasant odours. The base of the trough should then be raised 10 to 30 cm, with the front 50 cm above this. If too much room is allowed, wastage can increase as the objective is to allow the stock to feed and then move back.

Farm office and staff quarters

Maintaining good farm records is much easier in a farm office. An area at home or in the dairy shed should be dedicated to keeping records. It must have a desk and good lighting. It must be a quiet place to set up the office files (preferably in a filing cabinet) and computer and office supplies. The how and when of keeping farm records depends on the person recording them and this is discussed in more detail in Section 7.1 in Chapter 7. Computers are very convenient but require money to purchase and skills to operate efficiently. They also need to be kept cool and dry. Record keeping should be given as high a priority as other farming activities. Suitable chairs and tables should be included for business meetings with service providers and other farm-related visitors.

Farm staff should be provided with space to eat and relax when off duty. This could include a shower and toilet, food preparation area, and storage for their work clothes.

4.3.4 Cooling cows

Effect of altitude on air temperature

The air temperature decreases 6.5°C for every 1000 m increase in altitude above sea level.

Garden hose

The easiest method is to hose down the cow for several minutes. Water should be applied to the head and back of the animal with enough applied so it runs off their back and down their sides.

Sprinklers

Sprinklers need to be suspended 2.3 m from the ground above the feed troughs with the water directed to the back of cows. The droplet size should be medium to large, depending on the humidity. A filter should be installed at the beginning of the waterline and the adjustable sprinkler nozzles should be easily removed for cleaning.

Cows should be sprinkled for 1 to 3 min, applying 1 to 2 mm of water per 15 min cycle, followed by a 5 to 10 min period when the sprinklers are turned off, to allow the ventilation (usually aided by fans) to dry off the cows' coats. The pipe size depends on the length and area of the shed to be sprinkled, the number of sprinklers and the flow rate. Use 32 mm diameter pipe for up to 30 m length or 51 mm diameter pipe for 60 to 150 m length. Nozzles should be spaced at twice the radius of their throw, for example, every 2.4 m for nozzles with a 1.2 m radius. Water misters are usually not suitable for use in the humid tropics, as they will often increase the humidity and reduce the ability of the air to cool the cow.

Cooling fans

Providing a fan increases the rate of cooling especially at low relative humidities. A simple electric fan, that can be moved around the shed to be strategically placed behind any cow, is an effective method to increase the cooling rate of heat stressed cows after being hosed down. These are also effective in humid areas, so long as the sprinklers are thoroughly wetting the cows.

Permanent ceiling fans can be arranged in many ways. A 0.5 horse power, 0.91 m diameter fan rated at 5–6 cu m/min will blow a distance of 9 m, while a 1.0 horse power, 1.21 m diameter fan rated at 9–10 cu m/min will blow a distance of 12 m. The direction of the fans should be with the prevailing wind. In wide sheds, the side-by-side spacing width of 0.9 m fans should be ~6 m, whereas 1.2 m diameter fans should be spaced 9 m apart. They should be positioned ~2 to 2.2 m above the floor. Fans should be tilted so they blow down to the floor directly under the next fan (~25–30° from the vertical).

Exhaust fans and roof vents, that do not require external power, may be worth considering to encourage greater heat removal through the roof.

Outdoor area for night-time

Air temperatures of less than 21°C for 3 to 6 h at night can minimise reductions in milk yield due to daily heat stress. Once the sun sets, outdoor areas are cooler than inside any shed. Therefore, it seems logical to allow the stock outside the shed at night. Cows will eat more during the cooler part of the day, so provisions must be made for adequate feeding space. It is better to provide the bulk of the forage in the evening, particularly during summer, so adequate feeding and watering spaces are important. Also, as it is easier to observe oestrus when cows are loose in a yard at night, permanent lighting is important. Even if they are restricted to tie stalls during the day, they can be untied for afternoon milking and re-tied after the morning milking.

Saudi Arabian dairy feedlots have outdoor areas, as do many Australian dairy feedlots. They are much less common in Asia because of the high rainfall and problems with mud. Where cows are frequently housed indoors, outdoor loafing areas should be an integral part of any shed design. Whether the floor needs to be concrete or brick or just plain dirt, is another issue. Free access to adequate watering space is essential.

A lounging area for the milking herd should:

- be located in a well-ventilated area
- be well drained
- have a dirt or gravel surface
- provide at least 5 to 10 m^2/cow
- incorporate water troughs
- incorporate troughs for feeding forages.

4.3.5 Other farm buildings

Farm machinery should be stored under cover. Such a facility could be used as a workshop for minor maintenance and servicing of machinery and also a storage area for bagged inorganic fertilisers.

A biogas plant can easily be constructed to better utilise shed effluent. The gas stored in a large plastic bag would then be piped into the kitchen for cooking or into the milking parlour to heat up water for cleaning the equipment. It can even be used to produce electricity from a generator, but this requires special equipment to pressurise the gas. The residue from the biogas plant is still a good source of organic fertiliser for growing forages.

4.3.6 Housing for other stock and milking parlour

As these topics are more relevant to herds larger than 10 milking cows, they have been presented in the following Chapter, at Sections 5.3.3 and 5.6 respectively.

4.4 Feed storage

Feeds should be stored in secure locations with easy access to the stock. Vermin control will be necessary. They should be located well away from waterways. Most feeds are sourced in bags, although larger farms may handle feeds such as by-products in bulk. In these cases, good road access for trucks is essential with the feed bunkers under cover. Such bunkers should have concrete floors and brick, concrete block or prefabricated or reinforced cement walls. For storing by-products with very low dry matter contents, soil ramps adjacent to these bunkers will aid discharge from trucks.

4.4.1 Storage of conserved forages

Forages can be conserved dry as hay or straw, or moist as silage. Hay or straw must be stored in weatherproof buildings and should not be stored in confined places if it is likely to be moist, where it can become mouldy or spontaneously ignite. Silage must be continually covered with plastic sheets to exclude any air before feeding out. Silages can be stored above ground in bunkers or below ground in pits. In-ground pits are best for long-term storage, but above-ground storage is preferred in areas with high rainfall but must allow drainage from the pit. Buns or stacks are the silage units without structural support. However, with their high surface area to volume ratios and plastic costs per tonne of silage and wastage rates, costs can be high. Bunkers or clamp silos are the most common storage systems, while individually covered bales or stretchable bags are often used in regions well serviced by silage-making contractors. The dimensions, particularly their width, should be determined before their construction, based on the predicted daily rate of removal from the silage stack.

Silages can produce a lot of liquid during storage, called leachate, which is very corrosive and lethal to aquatic life if it escapes into rivers or streams. The production of leachates can be reduced by incorporating dry material such as rice straw or rice bran at the base of the silage bunker.

4.4.2 Preparing concentrate mixtures

It is rare for milking cows to only be offered one type of concentrate, because of their high nutritive demands and the poor nutritive value of tropical forages. Concentrate mixtures usually contain a variety of energy and protein supplements together with minerals, usually macro minerals, and for high yielding cows, maybe even micro minerals, probiotics and other feed additives.

One very simple method of hand-mixing a complex concentrate mix for storage or immediate feeding is to spread each ingredient on top of each other in layers (even the mineral premixes), then collect the mixture into bags or buckets by shovelling it perpendicular to the floor. This method is ideally suited for dry concentrate ingredients, although it could be used each day for offering mixtures containing wet by-products such as cassava or soybean waste.

Molasses blocks are a good supplement, and can be used as carriers of non-protein N or minerals. The blocks are solid, fairly easy to make, transport and store.

Molasses intakes should not exceed 20% of the total intake, because at higher levels it will depress digestibility.

Many farmers in Asia mix dry concentrates with water to make a slurry, before feeding. There is little benefit in this practice in terms of encouraging stock to consume more or improve its utilisation.

If the basic ingredients of concentrate mixtures are stored, a space must be allocated to blend them into a formulation. Hand-mixing may be practical with small herds of stock, but a mechanical mixer is preferred to ensure complete blending of mixes for larger groups of animals. These formulations are either placed in top of forages or fed out following removal of residual forages from the feeding troughs. With large herds, however, a mixer wagon is a good investment as it ensures consistency of ration formulations. Mixer and feed-out wagons are even available for small milking herds of as few as 20 cows. Such mixtures are called total mixed rations (TMR), or partial mixed rations if cows also have access to grazed pasture.

4.5 Checklist for planning dairy sheds

Physical aspects

- Aspect of shed with respect of wind and sun.
- Flooring (material, slope).
- Rubber mats for floors.
- Wooden or metal uprights.
- Roof height.
- Roofing material (colour bond or tiles).
- Ensuring maximum natural ventilation.
- Feed troughs (cement, wood or other material).
- Water drinkers (automatic, trough plus float valve, trough plus hose).
- Location near feed growing areas (for effluent recycling).
- Design of shed to allow easy stock observation.
- Wide aisles for ease of carrying heavy loads.

Services

- Electricity (single or three-phase).
- Water (drinking, cooling, washing, sanitising).
- Water heating (liquid petroleum gas or fire).
- Water for washing down, taps and hose or floor flushing.

Environment

- Insulation for under roof.
- Sprinklers for cooling.

- Hose for washing down hot cows.
- Cooling fans (large overhead or small moveable).
- Fans for all milk cow areas or just for high yielding cows?
- Computerised system to initiate specific cooling cycles.
- Plant and protect trees, grass around sheds.

Effluent disposal

- Suitable gutters to keep floors clean.
- Effluent pit (sufficient capacity).
- One large pit for several sheds.
- Liquid effluent removal (pump and pipes).
- Solid effluent removal.
- Location of fodder growing areas for recycling.
- Area to dry manure and make bricks/cooking fuel.

Feeding facilities

- Separate feed storage areas.
- Bird, vermin and insect-proof.
- Silage pits.
- Bunkers for urea treated straws.
- Fresh or dry forage feeding.
- Hay shed for bulk storage of purchased hay/straw.
- Concentrate storage (bulk or bag).
- Chopper for processing roughages.
- Feed mixer (horizontal or vertical) for concentrate formulations.
- Stationary mixer wagon to process total mixed rations.
- Farm utility for transporting feed, stock and milk.
- Farm tractor for handling bulk feed.
- Floor space for hand mixing concentrates.

Milk harvesting

- Hand or machine milking.
- Separate milk harvesting area.
- Mobile mini milkers or milking plant.
- Vacuum lines above cows.
- Engine room for milk plant.
- Specific buckets/milk cans for each shed.
- Handling bulk milk (vat, milk can).
- Refrigerated milk vat in each shed or for entire complex.
- Other milk processing equipment (pasteurising, packaging).
- Specific equipment for cleaning/sanitising milk-handling equipment.

Adult cows

- Tie stalls or free stalls.
- Adequate feeding space per cow.
- Access to drinking water.
- Level of milk production.
- Grouping cows on stage of lactation.
- Halter and rope or neck yoke.
- Separate large yard for heat detection and night cooling.
- Handling yards and crush for veterinary attention (see Figure 4.1).
- AI or natural mating.
- Mating yard for bulls.

Young stock

- Separate milk rearing shed.
- Separate weaned young stock shed.

Figure 4.1: Smallholder farmers, often with limited resources, have to be innovative with their farm facilities. This Kenyan dairy farmer, with just 4 milking cows, generates a high proportion of his dairy enterprise income from the sale of pregnant dairy heifers. He obviously needs a strong cattle race in which to inseminate and provide health care. He considers this adequate.

- Suitable capacity for 25–30% replacement rate in same shed.
- Design of calf-rearing pens (above ground or deep litter).
- Location of milk-fed calves (away from adult stock) (see Figure 4.2).
- Material for calf pens (bamboo, wood, metal).
- Area for mixing milk replacers and calf concentrates.
- Hot water for cleaning equipment.
- Feed scales for weighing all feeds.
- Livestock scales for weighing stock.
- Hospital/isolation pen for infectious diseases.

Personnel

- Office and suitable furniture.
- Computer facilities.
- 'Smoko' or tearoom.
- Living quarters (if overnight work is required).
- Lockable storage of veterinary drugs and valuable equipment.

Figure 4.2: Calf hutches provide an excellent way to rear replacement heifers outside. Such a facility would be equally ideal for a smallholder farmer or a large-scale dairy feedlot farmer.

5
Physical features on large-scale farms

This chapter presents an understanding of the important physical resources necessary for large-scale tropical dairy systems to be viable.

The following physical features are set out below for a large-scale dairy farmer milking a herd say of 100 or more dairy cows, as well as rearing their own heifer replacements. If the ultimate plan for the farmer is to milk more (say 1000 cows) or fewer (say 80 cows) stock, then the size of the physical features need to be adjusted in proportion to the number of total dairy stock. This chapter is similar in layout and detail to Chapter 4, because the principles for the design and layout of the dairy farm are not dissimilar for both smallholder and large-scale farms. This chapter should then be read in conjunction with Chapter 4, particularly as it will not duplicate the following sections:

- other facilities in cowsheds (4.3.3)
- cooling cows (4.3.4)
- other farm buildings (4.3.5)
- feed storage (4.4)
- checklist for planning dairy sheds (4.5).

5.1 Land

5.1.1 Location

See Section 4.1.1 for discussions for smallholder farms.

5.1.2 Forage production area

The key principles are outlined in Section 4.1.2 for smallholder farms. The following refers specifically to the layout of larger-scale farms.

Area of forage grown

For a 500-cow farm where all the stock's forages are to be produced on-farm, the farmer should aim to utilise a total of 60 to 80 ha land for forage production. This recommendation is for a farmer in the humid tropics growing Napier grass as the primary forage source. This will obviously vary with the location and preferred forage species, but because of its relatively high yield, the required area is likely to be larger in regions where forage species other than Napier grass are grown.

Area available for grazing stock

The optimal stocking capacities in Section 4.1.2 are all based on the forage being mechanically harvested and fed to dairy stock housed in sheds. Additional land can be developed for grazing for some classes of stock, such as growing heifers or dry cows. However, combining grazing with 'cut and carry' forage feeding systems in the tropics requires an additional set of skills, namely grazing high yielding cows in the humid tropics. If land is available, grazing should be preferentially offered to growing heifers and dry cows, as they do not require twice daily attention as do milking cows and milk-fed calves.

In some areas, milking cows may be provided with large areas of pastures for grazing as well as indoor feeding facilities. The indoor feeding facilities are then called feedpads or feedpad sheds. If the feed offered indoors is mixed and delivered using a mixer wagon, it would then be called a partial mixed ration (PMR), rather than a total mixed ration (TMR) that is fed to stock living inside sheds.

Other aspects of forage production and utilisation

Fresh forages fed to large milking herds are usually harvested mechanically, although they may be harvested by hand using large teams of farm staff. The forages are then transported to the cowshed, either by vehicle if distant, or by hand or hand-pushed trailer if close by. They may be wilted before being fed out to reduce moisture content, in which case a series of wilting racks needs to be constructed close to the cowshed. Finally, they should be chopped, either by hand or machine, before feeding out.

Large-scale harvesting of forages is best undertaken using a forage harvester that chops the forage before directing it into a large wagon for transport to the feed shed.

5.1.3 Effluent storage and management

From Section 4.1.3, a 500-cow (plus replacements) farm thus produces 16 to 18 t fresh manure and 7000 to 10 000 L urine each day and this would amount to a total volume of 15 to 20 m^3/d, excluding any rainfall run-off.

Any farm with more than say 100 milking cows is too big to rely entirely on physical labour to collect and distribute the shed effluent. Therefore, to minimise labour, such a farm would require a large storage pond plus manure pump and distribution system. The size of the storage pond would depend on the annual rainfall, the ability to divert run-off water to other disposal systems, the daily input from the dairy shed and the frequency of applying the effluent to the forage production area. However, it would have to have a capacity of at least 50 m^3 to provide for several days' storage.

The depth to groundwater under the soil surface is important as this will limit the pond depth and so increase its required surface area. The base of the pond should be at least 1 m above the highest seasonal watertable. With large dairy farms, the permeability of the soil is particularly important, as this will influence any seepage to groundwater supplies.

Effluent solids removal is also an option, with the solids being spread on grass or forage, or being sold and the liquid used for irrigation. Large-scale bio digestion and electricity production is also an option worth considering with solid waste used as fertiliser.

5.1.4 Water supplies

It is essential that cows are provided with adequate supplies of good quality drinking water. If this is not readily available year-round, then it can be supplemented with a system for collecting the rainwater during the rainy season for storage until required. Throughout the humid tropics, greater attention needs to be given to ensuring all sheds and houses have gutters and downpipes to facilitate the collection and storage of rainwater for stock drinking purposes.

Lactating dairy cows in the tropics require 60 to 70 L of water each day for maintenance, plus an extra 4 to 5 L for each litre of milk produced. Water requirements rise with air temperature and an increase of 4°C will increase water requirements by 6 to 7 L/day per L milk produced. High yielding milking cows can drink 100 to 150 L water/day during the hot season, so a 500-cow herd would require up to 75 000 L/d to ensure limited water intakes do not restrict appetite. In areas with lengthy dry seasons, very deep underground bores may be the only reliable source of stock water. The high cost of accessing this water often leads to shortages in drinking water, which can reduce feed intakes, accentuating the problems of reduced milk yields arising from depressed nutrient intakes.

5.2 Buildings

Cowsheds need to be sited to ensure a balance between dairy production, livestock health, environmental protection and long-term farm sustainability, and at the same time have minimal impact on neighbours.

There are many issues that must be considered when selecting a site for cowsheds on large farm. These include:

- access to the milking shed, feed supplies and farm laneways for stock movement
- access for employees and vehicles
- access to a breeze to cool cattle
- being of a suitable size, height and distance apart to allow breeze for ventilation and cooling
- being suitably placed for maximum biosecurity
- highland areas, because the air temperature decreases 6.5°C for every 1000 m increase in altitude above sea level
- being close to areas potentially destined for large-scale industrial development, because this can change the microclimate of the farm over several years
- expansion opportunities and staged development programs
- access to sufficient water of acceptable quality
- slopes and other topography features
- visibility from milking shed or farm residences
- any existing site services, such as channels, drains, electricity
- existing vegetation, particularly native vegetation
- proximity to waterways, dams and streams
- existing buildings, particularly farm residences
- prominence in the landscape, views into the site and general aesthetics and amenities
- provision of shade in the absence of trees on the farm
- boundaries and easements
- flooding impacts
- groundwater impacts and consequences
- uncontaminated stormwater control or diversion
- ease of harvest of manure and reuse of wastewater and nutrients
- prevailing winds in relation to odours and farm and neighbouring residences
- potential conflicts with future development of urban areas, residences, land zoning
- proximity to archaeological or heritage sites
- proximity to threatened or endangered species or ecological communities
- proximity to underground telecommunications, water reticulation or electrical or gas transmission conduits.

5.2.1 Topography

The natural drainage regime of the area must be considered to facilitate the diversion of uncontaminated storm run-off, the removal and transfer of wastewater to an effluent storage, the reuse of wastewater and the drainage of contaminated run-off from the feed storage area.

Slope creation on a cowshed site relies on having sufficient soil depth to accommodate the cut and fill requirements necessary for undertaking earthworks. This applies particularly to areas where holding ponds are required.

The sheds should be formed well above the natural surface level to promote drainage, increase air movement to decrease odours and to discourage insects. The cost of sourcing and moving sufficient suitable soil needs to be taken into account. Buildings and works should not be located on steep slopes, greater than 20%.

5.2.2 Flooding

Before determining a potential shed site, a property floodplain overlay should be obtained from relevant local or state government agencies to clearly identify potential flood-prone areas. This should avoid any adverse impacts during significant rainfall events and periods of prolonged rainfall.

The sheds should be located above the 1-in-100-year flood level and all earthworks should be designed to avoid offsite impact of floodwater discharge either through funnelling or backwater effects. The tops of any farm dams should also be above this flood level, but not necessarily any irrigation recycling sumps used to shandy effluent.

The natural drainage regime of the area must be considered so that uncontaminated rainfall run-off can be easily diverted away from the effluent storage. Gutters and downpipes leading to drains or water storages will allow more effective control of such rainwater.

5.2.3 Groundwater impacts

Groundwater investigations are also important when designing sheds and determining potential sites for their construction. It is often advisable to dig check wells at various localities and monitor them over a 24 hr period to see if they fill with water.

Cowsheds and associated areas, such as cattle loafing and silage bunkers, have the potential to generate high levels of nutrients which in turn pose significant risks to both surface and groundwater supplies through surface run-off and leaching. These risks need to be identified and managed during the development phases of the farm.

5.2.4 Shed slope

A fundamental design aspect of any large shed site is ensuring the floor slope is adequately determined and constructed to optimise the proposed site's natural topography. This will also avoid unnecessary soil importation, which will add significant cost to the construction.

The overall slope of the floors and associated loafing areas is important to:

- effectively remove effluent, fibrous material and sand from alleyways by optimising flood washing systems performance
- enable sufficient drainage from the site to remove water quickly following heavy rainfall events

- prevent unnecessary pooling of water creating odour and boggy areas
- enhance cattle flow and prevent potential flat slippery areas
- utilise the site's topography and natural drainage features
- eliminate tractor and mixer wagon slip when feeding out
- allow sufficient turning space for vehicles.

5.2.5 Soils

Soils need to be evaluated to assess their suitability for constructing the sheds and effluent storages. The engineered earth must be strong enough to carry the heavy loads of machinery, building foundations and, of course, stock. Clay dominant soils will be required and sandy soils avoided. Soil investigations and permeability tests should be undertaken to avoid costly mistakes and risk to the surrounding environments.

During shed construction, better compaction can be achieved by removing and stockpiling the topsoil to use later for landscaping or covering exposed soils (with a minimum layer of 100 mm to protect against erosion).

Insufficient soil investigations for shed sites can lead to:

- constant maintenance and repairs of high traffic areas
- surface cracking
- groundwater pollution from nutrient leaching
- seepage into effluent storages.

5.2.6 Access for stock and vehicles

The sheds should be in a central location to minimise vehicle movement and provide for orderly stock management and effluent collection. A tyre wash should be considered for all vehicles entering the farm as part of the biosecurity system.

The physical dimensions of the laneways, races, gates, entrances and exits should be designed for ease of stock access and movement. Vehicles require a minimum of 3.7 m width for easy access. A compromise is usually necessary between very wide laneways, which occupy land and reduce stock control, and narrow laneways with a funnelling effect that can contribute to problems with lameness. Factors to consider include:

- structures should take advantage of the social behaviour and natural movement of cows
- fences and gate should have no protrusions to hinder stock movement
- laneways should be a minimum of 4 m wide which should be increased to 5.5 m for herds of more than 120 cows. Laneways 8–10 m wide are now commonly used on farms with herds in excess of 1000 cows. Very wide laneways can reduce control of stock movement
- allow stock to enter the sheds without having to change direction by more than 90°. Use rubber matting on high traffic and turning areas to reduce hoof erosion
- all laneways, access and stocked areas should be well drained with the run-off directed to drains leading to the shed effluent system.

The shed design should also allow for all-weather access of vehicles with:

- minimum width for laneways of 4 m
- feed lanes should be 4.5 to 6 m wide to allow for easy tractor and feed-out wagon access
- entry and exit points, as well as turning areas for cleaning and feeding out, should be wide enough (8–10 m) to allow free flow of stock and vehicles
- room for distribution of feed on the feeding area
- room for cleaning operations
- movement of trucks to deliver feed on site
- access from a main or secondary road
- sufficient room for large trucks requiring high clearance
- adequate area for collection and stockpiling of solid wastes and residues from effluent storages and solids traps.

5.2.7 Constructing and maintaining external laneways

Routes for laneways should permit easy cow flow and, if necessary, allow for herd expansion. Grass and topsoil should be removed and the prepared area well compacted, above the watertable and free from pugging. Additional materials, such as pit or river gravel, may be necessary to elevate or stabilise the surface in preparation for track construction. This layer should also be well compacted.

First, construct a sound base that can support the top surface without moving, wetting or breaking up:

- using coarse material with a clay content of 15–30%. Do not use soft clay
- applying the material in 150 mm layers and compact well between layers with a roller at slow speed. Use a power grader with an experienced driver, or alternatively, a tractor mounted blade and a pneumatic-tyred roller or loaded vehicle with high-pressure tyres
- developing a camber or slope across the track that will shed water and be comfortable for cows, using a slope of say 1 in 10.

Second, construct the top (wearing) surface suitable for cows to walk on, ensuring:

- it has a minimum depth of 50 mm, with 100 to 150 mm preferred
- it does not contain stones which can cut hooves, or free uneven material that can cause bruising
- if coarse gravel is used, it is well rounded gravel of less than 25 mm diameter, with 15–30% clay content to allow adequate compaction, binding, wear resistance and smoothness
- large stones are avoided as these get kicked out of the track, leaving the site susceptible to water ponding and hoof damage
- the surface is sealed and the camber finished to prevent moisture draining into the base. The addition of 3% cement before compaction assists sealing
- compaction is essential as cows are not able to compact tracks adequately.

When maintaining laneways:

- seek out sections that are breaking up to identify and rectify any underlying causes
- drainage may need to be improved or the top surface may require additional compaction
- repair sections closest to the dairy first, as these are ones most frequently used
- remove stones from these sections and make sure they are well compacted, appropriately crowned and well drained. Cows cannot adequately compact these wearing surfaces
- if problems exist with cow flow, that section of the track may need drainage, repairing, crowning, maybe widening or even rerouting.

It is important to avoid right-angle bends in laneways as cows tend to bunch up, slowing down movement and increasing effluent dropped. Manure in laneways is less likely to enter the effluent system to be redistributed around the farm, thus losing its potential as a useful soil nutrient. Good cow flow minimises the time cows spend on laneways, hence the lost effluent. Allowing cows to move at their own pace will also reduce lameness and fouling of laneways.

5.2.8 Height and orientation

The height of any shed should be accessible to machinery used for cleaning and feeding, with a minimum recommended ridge height being 9 m and eave height of 5 m with central roof ventilation. It should also promote sufficient ventilation and penetration of sunlight. As the positioning of eaves and roof canopies will affect shading and sunlight, orientation is important. Winter sunlight provides additional warmth for the stock, helps to dry the feedpad surface and reduces the incidence of diseases. Summer shade reduces heat stress.

The sheds can be orientated to either maximise or minimise exposure to climatic elements such as prevailing winds, sunlight and rainfall. Extreme weather conditions can affect all feedpad operations from time to time, regardless of geographic orientation. Orienting the feedpad perpendicular to the summer and autumn prevailing winds assists in cooling the cattle. For a covered or shaded shed, a north–south orientation promotes drying because the shade moves across the pad during the day. Dry shed conditions are important as wet areas can predispose the stock to mastitis. However, since a north–south roof or shade orientation does not provide any less area under shade compared to an east–west one, covered free stall sheds should be orientated east–west.

5.2.9 Drainage

Effective drainage is important for all weather access, collection of contaminated run-off and diversion of uncontaminated stormwater. Raised floors will promote natural drainage. The drainage system should be designed to handle shed run-off from a 1-in-20-year 24 h storm event. Additional water from all surfaces, as well as washing systems and trough spillages, should be included in any storage volume calculations.

The drainage system should incorporate:

- drains or diversion banks
- sedimentation basin to remove solids from liquid effluent
- catch drains to carry storm run-off and effluent, with minimum slope of 0.5%.

It is important to minimise the necessary capacity of effluent storage requirements. Rainfall contaminated by faeces, urine or residue feed should obviously be directed into effluent systems. However, uncontaminated rainfall run-off (from clean surfaces such as roofs) should be diverted to the natural drainage system.

Run-off from livestock areas contains relatively high levels of nutrients, salts, chemicals, debris, pathogens and oxygen demanding organic matter so must be directed to effluent re-use systems (see Figure 5.1a to d). Wherever changes are made to cow flow, provision should be made for the diversion of contaminated run-off to effluent systems.

5.3 Specific details of buildings

5.3.1 Free stall cubicles and sheds

A free stall shed is essentially a feedpad with the addition of specific bedding areas where the stock can lie down. It is generally a covered shed and may include a loafing area for cattle to also be loose housed where they can stand, ruminate or idle. When well-designed and managed, free stalls provide the ideal system for intensively managing dairy cows off-pasture as each animal is provided with a specific place to rest, their management (feeding, cleaning and relaxing) is potentially optimised and the system can operate efficiently with minimal labour. However, they are expensive to construct and can become very unprofitable if stock suffer from poor welfare, animal health and milk quality due to poor feeding, cow comfort and herd management.

Free stalls are individual cow bedding cubicles where partitions orientate stock for comfort and sanitation, providing each cow with a dry and comfortable place to lie down and rest and ruminate. Free stall sheds should have one stall for each lactating cow. Some farmers provide additional stalls to allow for herd growth and to provide areas for subordinate animals to move away from more aggressive herd mates.

Stalls must allow enough room for the largest cow to freely enter the stall, lie down, rest comfortably and easily get to her feet to exit the stall. To do this, stalls should take into account the cow's normal desire to rest facing uphill slightly and

Figure 5.1: Managing cow effluent is a major issue on any large dairy feedlot farm. This series of pictures shows various ways it can be addressed. (a) The floor scraper is a chain driven device that operates 24/7. It is the most popular system to remove shed effluent even though it cannot completely clean the cow walkways. The shed effluent is directed towards a 'solids separator' facility to reduce the liquid content sufficiently for it to be handled by machinery to transport and spread it on the forage-producing areas of the farm or to nearby contract forage-producing farms. (b) This effluent tank is towed down the cow walkways while the cows are at milking

Figure 5.1: (Continued)

or otherwise locked out of the free stall shed. It sucks up the effluent for further processing before distribution on the forage-producing paddocks. It would produce a cleaner walkway surface than the floor scraper system. (c) The flushing system provides the best and most efficient cleaning system but there has to be a series of drainage pits to reduce the water content to allow the handling of the solids. Either that or the entire effluent solution is applied by flood irrigation. (d) The water flushing system requires large volumes of fast-moving water, hence the water towers shown in this picture. It is obviously better suited to high rainfall areas.

change resting positions or stretch while lying down. In addition, there needs to be sufficient room at the front of the stall as cows need to lunge forward to lift their hindquarters first when rising.

The shed layout should provide a flight route for submissive cows. In pens with rows of free stalls, there should be at least two cross-alleys, even in small groups of eight cows. Per 20 cows, there should be at least one cross-alley as a cow should never have to walk past more than 10 stalls to reach a cross-alley.

Cubicles or stalls can be arranged in a single row or in more than one row with a central feeding alley or with feeding alleys along the sidewalls. The cubicles can be arranged with cows facing one another (head-to-head) or the other way around (tail-to-tail). With the tail-to-tail arrangement, a central cow alley, 2.2 m wide between the cubicles is needed. If the cubicles are head-to-head, two cow alleys behind each row are necessary. Usually one of these alleys is combined with the feed alley. Free stalls are usually laid out in modules with crossovers providing access to the feeding alley. These can provide multiple routes between cubicles and feeding area and so minimise the adverse effects that dominant stock can have on the eating behaviour of submissive stock. Some free stall sheds have self-catching lockable feeding headstalls along the feed bunk to allow animals to be caught for veterinary attention, insemination or even locked away from the feed.

The latest information in relation to stall dimensions must be checked before any construction beginning. The following is a guide only.

Stall dimensions should be based on the largest 25% of the herd to allow for increase in cow size through improved feeding and genetics over time. They should also provide for adequate lying down as well as necessary forward and sideways lunging to stand. Typical Friesians require ~240 cm long × 120 cm wide lying space with a further 60 cm forward lunging to allow for normal standing behaviour. Forward lunging space can be shared where two rows of stalls face head to head.

The size of the cubicles depends on cow size. For cows:

- weighing 400–500 kg, they should be 104–109 cm wide and 198–208 cm long
- weighing 500–590 kg, they should be 109–114 cm wide and 208–218 cm long
- weighing 590–680 kg, they should be 114–122 cm wide and 229–244 cm long.

Stalls that are too long or wide allow the animal to move forward or at an angle, in which case faeces and urine can be deposited within the stall and not in the alleyway. To further prevent cows from soiling the cubicles, shoulder, breast and neck rails are needed to force cows backwards when they stand up. The distance of the adjustable shoulder rail to the back of the cubicle, measured diagonally, should be ~1.8 m, and the height to the cubicle floor may vary between 0.9 and 1.05 m.

The stall kerb separates the stall area from manure in the walkway. It should be high enough to prevent manure from entering the stalls, but low enough to allow cows to enter and exit the stalls easily. Recommended maximum curb heights are 20 cm or, if a mattress or mat is used, 30 cm.

The ideal lying surface is soft, absorbs moisture and does not promote the growth of bacteria. When cows are forced to lie on hard surfaces, they do not lie down for long, are more unsettled and may develop knee and hock lesions and swelling, acidosis or laminitis. The stall base is usually made of concrete or compacted fill, which is then covered with bedding, or even a mattress filled with recycled rubber (or crumb, foam, gel or water). Alternatively, rubber mats can be used with bedding. All base types need loose bedding material on top for further cushioning, moisture absorption and to reduce friction. If the stall base provides good cushioning, less bedding is needed on top.

The simplest and most effective bedding is concrete with a deep layer of sand, but it needs care to maintain the surface. Cows often push sand around, but a cement base, compared to an earthen base, can reduce this. Sand bedding can cause problems with mechanical or liquid manure handling systems because it fills up storage tanks and is very abrasive, damaging equipment such as manure pumps.

A concrete foundation with a disposable bedding of chopped straw, sawdust, wood shavings or crushed corncobs is more common in Europe, as rice hulls are not readily available. Rice hulls makes ideal bedding for free stalls. Hard surfaces should have a slope of at least 1% so that urine will drain into the cow alleys.

Hard rubber mats provide little cushioning, particularly if very thin, and they may be slippery. Soft rubber mats provide similar features as mattresses. Attachment methods, surface texture and compaction of the mat or mattress material are all issues to consider when selecting and installing mats or mattresses. Bedding is required on top of mattresses and mats to help to maintain clean dry conditions.

Manure and wet bedding should be removed and replaced with dry bedding material at least once per day. Cleaning should be frequent enough to keep the back of the stall clean because this is where the cow's udder and teats are in contact with the bedding when she lies down. Organic bedding (sawdust, straw, hay, composted manure, rice hulls) should be added daily, especially on mattresses and rubber mats, as it is hard to keep bedding on these surfaces.

Dirty cow alleys will result in dirty beds and udders, weakened hoof horn and potential mastitis. Cow and feed alleys should be kept clean by manual scraping, automatic scrapers or flood washing. Although cows can still be in their stalls, it is better to time flood washing when they are away from the shed.

Mechanical scrapers are probably the most common floor-cleaning equipment in free stall sheds but they still leave a film of cow manure on the floor, thus continually exposing stock to pathogens. One operator in Vietnam commented that in such a shed, lameness and mastitis levels increased each summer, when sprinkler systems were in common use to cool the cows, because the additional water on the floor (from the sprinklers) increased the degree of exposure to these pathogens. Therefore, ideally the floors in sheds that use mechanical scrapers should also be hosed down every day to reduce this pathogen load.

The free stall environment should be made safe for the stock through ensuring they cannot put their heads through gates and fences or get stuck under stall divisions and barriers. There should be no projections, such as broken boards or rails or rough, sharp edges on the concrete. Rails should be strong enough not to break when cows lean on them. Walking surfaces should be grooved to minimise slips and falls and so encourage normal oestrous activity. Cattle crushes and head gates should be well designed to ensure that stock can be examined without fear of injury.

The free stall facility should be designed to ensure smooth and quiet cow flow. There should be no sudden changes from light to dark, reflections or drains across the cow alleys. Cows will move more smoothly along curved races, up a slight incline and where they have 'sure' footing. Gates can be muffled by attaching rubber strips to prevent excessive noise when closing. Yards must be designed for easy drafting of targeted cows. Stock should only be moved around using 'flappers' (leather strips attached to a cane) rather than using wooden or metal pickets or pipes. Excessive twisting of an animal's tail is unacceptable and electric prods should only be used in emergencies.

Monitoring free stall use

The cow is the final inspector of free stalls and if cows are not successfully and regularly using them, or they are dirty and show signs of injury, action is required. There are a variety of ways to monitor the cows' use of free stalls and free stall sheds, such as:

- Do cows appear comfortable when standing or lying? If not, stall dimensions and bedding may need attention.
- Do cows have to push, bang and/or bump against stall components to lie down, get up or change positions?
- Do cows lie backwards or sideways in the stalls or in the alleys?
- Do cows stand half-in or half-out of the stalls? This is called 'perching'. This can occur when the stalls are too short, the neck rail is too far back or when the stalls are otherwise uncomfortable.
- Do cows stand in the stalls at an angle? This indicates the stalls are too wide.
- Are all stalls used equally? If not, there would be a reason why some are not chosen by the cows.
- When cows normally rest (between 10 p.m. and 4 a.m.) are more than 20 to 30% of the herd standing in the stalls? If so, stall comfort may be questionable.
- Cows should lie down for at least 11 to 12 hr each day.
- Are cows' udders, tails or hindquarters dirty? This could indicate dirty bedding but may also be due to low fibre diets and very loose manure.
- Are there patches of rubbed-off hair or visible injuries to hocks and knees? These are signs that cows rub excessively on stall floors, partitions or neck rails when rising or lying down.

- Are cows walking very slowly, or timidly, with rear feet spread wide? This could be a sign of poor traction or laminitis.
- Are some cows slipping and falling in the shed? This could also be a sign of poor traction.
- Do more than 20% of the cows defecate in the milking parlour? This could indicate discomfort or uneasiness in the free stall shed.
- Are cows bellowing excessively or exhibiting abnormal behaviour?
- The comfort of the stall bedding can be assessed by the:
  - wet knee test, which involves kneeling in the stall for 10 s and if the knee is wet, the stall bedding is not dry enough
  - drop knee test, which involves crouching and then dropping to your knees in the stall. Any pain reaction in your knees will quickly tell you how truly comfortable the stalls are.

5.3.2 Facilities and equipment for additional health care

To ensure good health care, sufficient health facilities are needed and cowsheds should include additional stalls for such purposes. These and other stock handling aids are included in the following list.

- *Treatment area*, for confining animals on heat, artificial insemination, routine health checks, pregnancy diagnosis, lameness and examining sick cows. As animals are usually separated when they leave the milking parlour, it should be located close by. The width of this treatment area should be at least 0.7 m per cow and the length 3 m. Farmers usually find it convenient to store equipment in a specific veterinary drug box in this treatment area for treating hoof problems, for trimming hooves, taking blood samples etc. There should be one treatment stall for every 20 cow stalls with a minimum of two.
- *Separation area or hospital pens*, to treat sick cows properly and prevent the spread of disease. It should be located close to the milking parlour. Drinking water should be available, with concrete floor and gutters to allow for frequent cleaning and sanitising. There should be one hospital stall per 50 stalls with a minimum of two. With tie stalls, there is little need for a special separation area.
- *Calving area*, to permit proper attention at this critical time. With loose housing, cows may need to be tied up in stalls and should calve down away from the milking herd and close to the calf pens. As with the separation area, ease and thoroughness of cleaning and sanitation are key features. There should be one calving stall per 30–50 stalls with a minimum of two.
- *Separate bull and mating pen*. Although farmers generally use AI to mate their cows, herd bulls are often kept to mate heifers or cows that have been unsuccessfully inseminated several times. Bulls should be kept separate from

the adult cow herd except when used for mating. A mating yard will allow for cows on heat to be separated from other stock for better control of the bull's activities. It is preferable to bring the cow on heat, once identified, to the bull and so provide room for mating activities in the bull pen rather than let the bull mate in the laneway or a free stall.

- *Cattle race, crush and/or head bales* or other ways of immobilising stock are important both for the stock (to reduce stress) and the staff (to reduce injury). A cattle race allows stock to be separated out and immobilised when requiring additional attention such as insemination, veterinary treatment or foot trimming. It is important that stock can be handled safely when individually separated from the herd. The farm should have a set of cattle yards that includes a race for AI, a crush for close veterinary attention and treatment and, if desired, a set of scales to routinely weigh stock. Special hoof-trimming crushes are available in which each foot can be lifted up separately and secured to aid trimming of the hooves. Cattle crushes and head gates should be well designed to ensure that stock can be examined without fear of injury. Head bales should be made of steel with sufficient robustness to hold a large unsettled animal.
- *Footbaths* should be available for routine hoof treatment, at least 2 to 3 m in length and 0.15 m deep. The width should be the same as the passage to prevent cows from bypassing the bath without using it. Double footbaths are better because they allow dirt to be washed off before treatment. The first bath tends to activate the dunking reflex that allows the solution in the second footbath to stay effective for longer. A solid platform of 3 m between footbaths will help shed off some of the wash solution from the first footbath. Emptying and refilling them should be quick and easy. Hulsen (2013) recommends that footbaths should contain 4% formalin or 38% commercial formaldehyde solution, although formalin is a human health risk. Copper sulphate can also be used in the second footbath. Each cow should walk through them once each week. The frequency should vary depending on hoof cleanliness and dryness and how much they show evidence of horn erosion or digital dermatitis. Placing footbaths in the raceway from the milking shed will ensure a more thorough foot treatment protocol.
- *Electric cow trainers.* These consist of wires positioned above stalls that carry electricity and are used to teach cows to step backwards when their back is arched before defecation and urination. The cows then alter their behaviour, which keeps them and their stalls cleaner. Electric cow trainers are, however, risk factors for silent heats, clinical mastitis, ketosis and culling, with herds not using cow trainers having less of these issues. In some poorly designed stalls, cows prefer to stand and experience the pain associated with lying and rising. When they are moved to better-designed stalls, the cows learn to use these trainers. Many farmers do not use them as they are more likely to unsettle temperamental animals.

- *Storage of veterinary drugs and other dairy equipment.* Veterinary drugs should be stored in a dark cupboard, and if required, inside a refrigerator. For security purposes, it should be lockable. Other dairy equipment, such as detergents and sanitisers for cleaning milking machines and equipment for trimming feet or dehorning calves, should also be stored in a specific location. Milk-harvesting equipment, such as buckets, sieves and milk cans, should be stored off the ground. Farmers should hang this equipment upside down to aid drying.
- *Farm office and staff quarters.* This is an essential part of any large dairy farm. This area would house the administration team and their equipment, such as office files and computers, and also provide a centralised meeting place for senior management and farm staff. It would seem logical that microwave ovens, for routinely monitoring the DM of TMRs and their ingredients, microscopes for monitoring semen quality and other laboratory equipment also be located in this area. Closed circuit TV (CCTV) cameras would be worth considering to provide for 24/7 observations of stock (in sheds for milking cows, calves and heifers) to monitor stock behaviour and their wellbeing. Suitable chairs and tables should be included for business meetings with service providers and other farm-related visitors. The farm staff should be provided with space to eat and relax when off duty. This could include a shower and toilet, food preparation area and storage for their work clothes.
- *Cooling cows.* The air temperature decreases 6.5°C for every 1000 m increase in altitude above sea level. Cooling cows is discussed in Chapter 4, Section 4.3.4 and also in Sections 5.3.4 and 5.5 below.

5.3.3 Housing for other stock

In addition to milking cows, farms have to house dry cows, milk-fed calves and weaned growing heifers and herd bulls. Some farms also rear non-dairy (dairy beef) stock such as heifers, bulls and steers that may or may not have originated from the milking herd.

The area needed to house young stock depends on:

- their number
- type of housing
- calving pattern of milking herd
- replacement percentages of milking herd
- division of herd into age groups.

A good management practice is to divide replacement heifers into the following four age groups:

1. 0–1 month-old milk-fed calves,
2. 1–5 month-old milk-fed and weaned calves,

3. 5–22 month-old weaned heifers and yearlings,
4. 22 months and older pregnant heifers.

Young milk-fed calves should be housed in clean and disinfected individual pens (0.6 × 1.3 m), separated from the cows. The pen partitions should be solid and smooth, 1.5 m in height. The raised floor could consist of wooden or metal slats 20 mm apart or the calves could be penned on a free draining deep litter system. Containing each milk-fed calf in a raised metal, individual cage provides the best housing as they are isolated from each other, live in a well-ventilated and clean pen, are easier to feed and water (as the farmer does not have to bend down to feed them) and, of most importance, it allows for much easier individual surveillance. Some farmers are concerned that young calves need to be kept in groups so they can learn to socialise, but this can happen post-weaning without any detrimental consequences.

Older calves can be housed in tie stalls (0.7 × 1.5 m) or preferably in group pens. Weaned heifers can be housed in tie stalls (0.8 to 1.0 m wide × 1.7 to 2.1 m long, depending on live weight) or preferably group pens or free stalls. Pregnant heifers can be housed with dry cows, although they will have different feed requirements.

If outside grazing areas are available, weaned calves and older heifers could be grazed as long as they can be provided with troughs for supplementary feeding when individual parasite treatment and shelter is required.

5.3.4 Other equipment

Such items will improve cow comfort and can become important factors beneficially influencing herd contentment in large sheds where there is likely to be increased competition between cows. There are considerably more antagonistic interactions within a herd of several hundred dairy cows than within a SHD herd, so having these sorts of equipment can reduce negative interactions.

- *Rotating cow brushes* are sometimes provided to allow cows to groom and scratch themselves. It may also reduce frustration or stress due to boredom. (Generally, cows need little behavioural enrichment because feeding, ruminating and resting occupy most of their time and they can rely on other cows for social stimulation.) Cows can be very vigorous in their use of brushes so this equipment needs to be quite robust. Some brushes automatically start rotating when an approaching cow is detected. As cows mostly use brushes to scratch their backs rather than their heads, they should be positioned accordingly.
- *Self-locking headstalls* are designed to restrain cows as they put their head down to eat from the feed bunk. Some free stall sheds have self-catching lockable feeding headstalls along the feed line to allow animals to be caught for veterinary

attention, insemination or even locked away from the feed. These then constrain the cow to facilitate closer observation or individual management. For example, some herd synchrony programs use coloured crayons to mark each potentially cycling cow (if she stands to be mated by another cow, the hair around the crayon is ruffled and easy to note) and these cows can be easily identified during feeding with the stock restricted in self-locking gates. In addition, they reduce aggressive interactions and displacement of socially subordinate cows while eating. These self-locking headstalls are not preferred by cows when they have an option. Despite this, when forced to use them, they have little impact on daily feed intakes, milk yields, levels of mastitis and signs of stress.

If heifers have never been exposed to these headstalls before entry onto the feedlot, some headstalls should be removed from the feeding area so the stock will readily consume the TMR, then reintroduced slowly over a few days or weeks so the stock will adapt to eating through them.

- *Cow showers* are designed to be activated either manually or automatically. In the latter, the showering and interval cycles are triggered only when the dual motion sensors detect an animal is present and the air temperature is above a specified threshold. As air temperatures increase, the interval time automatically decreases, thus giving animals more frequent shower and fan cycles to reduce heat stress. By using a high capacity, coarse droplet shower nozzle, enough water can be applied to fully wet the cows to the hide. Mist and fogging nozzles work by cooling the air around the cows and the disadvantage is that the mist can increase humidity and be easily blown away under windy conditions, or when used with fans. If a mist or fog builds up on the cow's hair coat, it can also trap a layer of air between the skin and the water, which holds in body heat. In comparison, soaker nozzles produce a coarse droplet spray, which penetrates the hair and wets the cow's hide.

  There should be a continuous flow of air over the backs of the cattle any time the cooling system is in operation. This causes the water to be evaporated, which takes the heat away from the cattle in the process. Fans can be controlled separately from the cooling system, and are set to operate continuously above a set temperature. It is worthwhile aiming some fans directly onto stock lying in the free stalls as well as having other fans directing air over their backs when they are eating.

  Normal recommendations are to shower the animals for a short period of time, 0.5 to 3 min, to soak the hide. After the shower shuts off, the water is evaporated from the cattle by fans blowing across their backs for 5 to 15 min, before repeating the shower cycle. Common locations for installing a shower cooling system are in the holding pen area, where cows are crowded together tightly, and at the feed bunk, but not so water will spray onto the feed or into the stall area.

5.4 Water and feeding management

5.4.1 Water supplies

Stock must be provided with sufficient drinking water at all times and the system should be able to supply at least 20 L/cow/hr to meet likely peak demand. The optimum temperature for drinking water is 15–17°C. Peak demand often coincides with feeding times. The pipe diameter needs to be at least 75 mm with an operating head of at least 10 m. A tank could be used for short-term supply in the event of a power failure. A backup water supply should be available to hold at least two days' peak requirements, in case of a breakdown or loss of normal water supply.

Upgrading the facilities can involve increasing the rate at which troughs fill or providing greater trough capacity. Cows can consume up to 20 L/min. Options involve upgrading pumps and water pipe diameters or increasing the number of watering points, particularly those close to the feeding area. Cows often drink a third of their daily water intake in the hour after morning milking on a hot day.

When depending on dams, water quality can become an issue because as dam levels drop, salt contents, bacterial loads and algal outbreaks all tend to increase. Grazing cows can tolerate reasonably high levels of salt in their drinking water, but tolerance levels drop as they are offered less pasture. Problems that limit water intake also limit feed intake, hence cow performance.

Water troughs should be well separated (but within 15 m) from feed troughs with the same water trough servicing adjacent pens. Provision should be made for water spillages, or leakages, to directly enter the drainage system and also for drainage control in the event of burst mains or a jammed float valve. Troughs should be located at the high point of a water line to reduce sediment and to facilitate purging air from the pipeline. Trough design should allow for regular, easy cleaning with a removable bung for complete drainage. In free stall sheds, they should be located at crossovers to reduce the incidence of stock blocking each other in alleys.

Each cow should be provided with 75 mm of linear watering space in free stall sheds while for circular water tanks, one watering space (60 cm of tank perimeter) should be available for every 15 to 20 cows. A water depth of 15 to 20 cm helps keep water cooler, fresh and easier to clean because less debris accumulates. The optimal trough height is 60–90 cm, from ground to the top of the trough. Watering points should be cleaned out at least weekly to remove any feed and other contaminants.

The passage around the drinkers should be at least 4 m wide, so when a cow is drinking other cows can easily pass by. At least 10% of the cows should be able to drink at the same time, but when the air temperature exceeds 30°C, 20% of the cows should be able to drink at the same time.

Access to a reliable supply of water of acceptable drinking quality is imperative. Watering systems may also be required for dust suppression and for sprinklers to cool cows. Although such run-off may be minimal, it should be directed into the effluent system.

If cows are tethered, it is possible to install individual cattle drinkers, with two animals sharing one drinker. To ensure water is freely provided, troughs need to have float valves rather than depend on a hose whenever the farmer decides to refill it. A single watering trough system can control water levels in up to 10 small troughs just by one float valve in a central reservoir with syphon pipes leading to all the troughs, although flow rates can be relatively slow.

5.4.2 Feed troughs and feeding strips/bunks

It is important that all stock can eat comfortably with minimal competition. Whether feeding into troughs or onto cement feeding strips, 70 cm feeding space should be allocated per mature cow. For a feeding strip or trough when cows eat from both sides of the strip, hence face each other to eat, this equates to a total feeding strip allocation of 35 cm/cow. Trough space can be reduced to 45 cm/head for 6 month-old cattle, or to 55–60 cm/head for stock 18 months old. Competition between animals is much less severe when complete rations are fed *ad lib* throughout the day, so trough lengths can be reduced, even down to 30 cm/head.

Ideally, troughs should be covered with a grate that directs the cow's head downwards and precludes the lifting of the head above the trough; however, this feature must be considered alongside the ease of filling the trough with feed. Self-locking head bales will reduce the incidence of cows throwing feed around as they lift their head, hence will reduce wastage. In the latest feedpad designs in the US, concrete feeding strips are covered with epoxy resin to reduce corrosion of concrete. Feeding strips should be positioned 75–150 mm above cow feet level with a nib wall to protect the feed from the effluent, 400–550 mm above the feeding strip. Trough width should be related to the reach of the animals which is up to 80 cm for mature dairy cows eating from the base of a trough 30 cm above ground level. Feed barrier walls wider than 15 cm can restrict how far stock can reach into troughs or onto feed strips.

Feed troughs and strips should be situated on the high side of the pen, running parallel with the contour to minimise pad drainage. They should also have smooth surfaces, as those without grooves or holes that can trap feed are easier to clean and help reduce build-up of waste feed, mould growth and unpleasant odours. The base of the trough should then be raised 10 to 30 cm, with the front 50 cm above this. If too much room is allowed, wastage can increase as the objective is to allow the stock to feed and then move back.

The height of any wall constructed to retain feed should accommodate the chute of the feeder wagon which should be higher than 60 cm, although it is wise to check this with the particular machine. In addition, there should be no uprights that might interfere with discharge from the feeding wagon.

5.4.3 Keeping cows out of the feed

The term 'feedbunk' is traditionally used in the US for feeding strips, but the practices listed below also apply to trough feeding.

- To maximise feed intakes and milk production in lofted stock, feedbunks should not be left empty for more than 2 or 3 h/d.
- When feeding time is limited to less than 8 hr/d, milk production can be reduced by 5–7% in mid lactation cows, and to an even greater extent in high producing cows, which are at or near peak lactation.
- Push up the feed regularly to encourage feeding and minimise sorting of ingredients.
- The feed barrier should not restrict cows' reach.
- It is important to provide sufficient feedbunk space so competition does not adversely affect feed intakes but not too much space so stock can throw feed around as they eat.
- As first-calf heifers tend to eat smaller meals more often than do older cows, separating them out in large herds might reduce competition and improve their performance.
- Feed surface is ideally 100 mm above feet level.
- Feed area should be under cover to prevent rain decreasing feed intakes.
- Remove residue fed and use for other groups, such as heifers or dry cows.
- Cattle consume most of their feed during the comfortable period of the day, that is in late evening in hot weather and during the middle of the day in cold weather. That is the time they should be offered the most.
- Cows are animals of habit that like routine. Once a schedule has been developed, stick to it and if changes are necessary, allow time for the cows to adapt and monitor feed intakes to decide if it was worthwhile.
- Ideally, feed residues should be restricted to 3% or less without impacting on milk production or cow health.

There is no easy answer to the problem of keeping cows out of the feed troughs, as it depends on the feeding system. Some ideas are shown below.

- To reduce bullying, cows can be separated into smaller groups, ensuring there is sufficient room at the feeding face.
- The trough length should be no more than 70 cm/cow if they are all feeding at once or 30 cm/cow if they are fed *ad lib.*
- Ensure the inside floor of the trough is 10 to 15 cm higher than the feedpad surface to provide easier access to the feed.
- Install an electric fence down the centre of the trough.

- Install pipe head bales. Although expensive, they prevent cows from climbing into or being pushed into troughs and also prevent them from lifting their heads out of the troughs, hence able to throw feed around.

5.5 Tunnel or assisted ventilation barns

The high relative humidity and high internal heat production, particularly in high yielding milking cows, create enormous problems with keeping cows cool enough so they maintain sufficient appetite for the high fibrous forages frequently offered them. Ideally in the humid tropics, large-scale dairy farms should be located in highland areas where air temperatures are lower, hence cows are less heat stressed. Evaporative cooling systems, where water droplets reduce air temperatures are often ineffective in the humid tropics because the air already contains too much moisture (as humidity). Thus cooling shed air has limitations in humid environments due to the lack of additional moisture holding capacity of nearly saturated air. As long as the cow's skin is so wet that applied water runs off the coat and there is sufficient air movement, her cutaneous evaporative system can cool the cow up to a point. Computer-controlled systems that spray large water droplets then utilise fans to promote air movement are in occasional use in many large-scale dairy farms in Asia's humid tropics.

Relatively new cooling systems for large sheds are ventilation barns housing 200 or more milking cows. These are often called tunnel barns and are designed to rapidly increase air movement over the stock to effectively cool them without the need to apply additional water to reduce air temperature. The principles are twofold, namely increasing air velocity to cool the stock, while at the same time improving air exchange to reduce the build-up of manure gases, dust and pathogens. The rapid air movement also discourages flies and birds in these sheds. The keys to success of this system are an adequate number of properly sized and located axial-flow fans together with a shed layout that will maximise the entrance of fresh air then its unhindered flow, at cow height, throughout the shed. Many different types of assisted ventilation systems are available as tunnel ventilation sheds are becoming popular throughout the humid tropics.

The back of these sheds holds a batch of very large fans (150 cm in diameter) which blow air out of the shed. Sometimes the front of these sheds have a second batch of large fans sucking air into the shed. This negative pressure then sucks air longitudinally throughout the shed's entire length. As well as promoting rapid air movement over the cows, it exchanges barn air uniformly with fresh outside air.

The shed design is completely different to the more normal free stall sheds, in that the layout facilitates the movement along the shed with a low roof and enclosed sides incorporating large plastic curtains. The fans can be retrofitted into existing sheds but they require extreme internal modifications so as to encourage unhindered air movement over the stock. Because the fans are not usually operating 24/7, shed designs should promote sufficient natural air movement to remove the airborne pathogens and gases when they are not operating. Curtain sidewalls must then be provided to act as air inlets and outlets. They can be closed during fan operations to form the 'tunnel'. In addition, horizontal air baffles along the roof are necessary to redirect the moving air back towards cow level. If there are no fans at the front of the shed, it should be as open as possible so as to maximise air movement into the shed.

The computerised controller would monitor air temperature, wind speed and direction to decide on which mode of ventilation is most desirable. If switching from natural to tunnel ventilation, the fans could be activated with adjustments made to curtain sidewalls and baffled air stacks. Such computer-controlled banks of tunnel fans should be programmed to initially be activated say at 21°C air temperature then become fully operational at say 27°C. These operating threshold temperatures will obviously vary with humidity and targeted cow milk yields.

Standby power generators are normally needed on dairy farms to ensure uninterrupted electrical service to milk-harvesting and cooling equipment during power outages. This must also include backup power to the tunnel fans and their control systems. Alarm systems are also needed to alert farm staff if an electrical service is interrupted to the tunnel fans, together with an evacuation plan to remove cows from the shed if needed. At the very minimum, automatic curtain drops need to be in place in the event of mechanical power failure.

The net return on investment is a function of how well cows maintain milk production for an adequate number of days in a tunnel-ventilated shed each summer as compared to a naturally ventilated shelter. The additional cost of purchasing, installing, operating and maintaining the tunnel system must be offset by the sustained milk production in order for this investment to deliver a return. In addition to lost milk production, reduced conception rates, compromised growth rates of unborn calves and sub-optimal cow health should also be assessed when performing a complete economic analysis of the system. A large-scale dairy feedlot has recently installed a tunnel ventilation shed in the highlands of East Java in Indonesia (1200 m above seal level) and the operators have calculated, for cows producing 30 L/day, the milk response was an additional 3 L/day of milk.

To date, complete design systems for tunnel ventilation are not available (however, see Figure 5.2 for an example from North Vietnam). Pertinent questions that remain unanswered include:

- What is the practical length of a shed that can be tunnel ventilated?
- How can uniform air velocity in the shed be best maintained?
- What are the optimal settings for automatic controls?
- What is the cost/benefit ratio for tunnel ventilated barns?

Figure 5.2: The tunnel ventilation shed would be considered an essential for any large-scale dairy feedlot in the humid tropics. It comprises a batch of large fans, as in this North Vietnam dairy feedlot. This shed has a second batch of fans at the other end of the shed, with one batch blowing and the second batch sucking, to produce a rapid air movement over the stock inside. In the dry tropics, 'water curtains' (or rubber strips over the fan outlets with water dripping off them) can be used to reduce air temperature but this system is ineffective in the humid tropics. The performance of high yielding cows in the lowland humid tropics will always be constrained for many months of the year by suppressed appetites which will reduce milk yields and increase the number of inseminations per conception. No matter what they cost, the tunnel ventilation shed must be considered a good investment. It is possible to modify non-ventilated sheds through a series of curtains along the sidewalls and baffles to direct the moving air over the cows.

5.6 Milking parlour and automatic data recording

5.6.1 Milking parlour

A separate milking area has many advantages. It can be centrally located requiring only one supply of hot water (for cleaning animals and equipment). Daily hose down and weekly disinfecting will minimise environmental constraints to good milk quality. Furthermore, if farmers eventually decide to invest in machines for milking, one milking area minimises problems with low vacuum pressures in long air-lines. In addition, if a decision is made that cows are allowed outside to cool down at night, it seems logical to plan their movements around milking times.

Cows can be milked in a variety of ways. Hand milking is very common with smallholder farmers throughout Asia while bucket milking can be used for cows in tie stalls. Loose-housed cows are usually milked in a separate milking area that can be specifically designed for ease of cleaning and occasionally sterilising. Hot water is essential for thorough, twice daily cleaning of all milk-harvesting equipment. Ideally, milk-cooling equipment should be installed, unless there is close access to a milk collection centre.

Bucket milking is the simplest and least expensive system to install, but the milk must be carried to the cooler. This type of system is usually used in small herds and where cows are milked on a level floor of the shed or milking area. A separate vacuum line can be built around the entire shed if the cows are to be milked while in tie stalls. The labour of carrying milk to the cooler can be avoided by installing a transfer system, which consists of a small receiving tank (say 30 L), including a built-in filter, mounted on wheels so it can be moved around the shed. Pipeline milkers transport the milk directly from the cow's udder to the milk cooler and are usually installed in milking parlours where the milker stands below the level of the cows. Although they are expensive, they save backbreaking labour and are usually designed to be cleaned in place, a feature that not only saves labour but helps to ensure good sanitation.

Designs of milking parlours

Any type of parlour should have a high quality concrete floor and metal rails for durability and ease of cleaning. Walls are not required, but if included they should be at least plastered masonry walls. The pit where the milker stands should have a floor level 90 cm below that of the cows for the most comfortable working position. The number of stands is determined by the milking herd size and the allowable milking time of the herd, taking into account the time taken for the cows to eat the concentrate rations often fed in the milking parlour. Parlours can be of the following types:

- *Abreast parlour.* The milker and the cows share the same floor space. Cows can enter and leave individually. The stands should be 1 to 1.1 m wide when a bucket milker is used or when the cows are hand milked, while 0.7 to 0.8 m is adequate when a pipeline milking system is used. The width for the milker should be 0.6 to 0.8 m. The main drawback is the relatively long distance to walk between milking points, and the cows obstructing the milker.
- *Tandem parlour.* This allows for individual care of the cows but has larger space requirements.
- *Walk through parlour.* Cows enter and leave in batches. Their narrower width can be an advantage if fitted into an existing building.
- *Herringbone parlour.* This provides a compact working area and allows feeders to be fixed to the side walls. Three or four stands on each side should suffice for up to 80 milking cows. Its popularity is due to its simplicity and high capacity

(measured as cows milked per man hour). However, the risk of cows kicking the milker is greater than in parlours where the milker stands beside the cow. The large ones are generally equipped with rapid exit gates in which the entire side of the milking area opens up once milking is finished.

- *Rotary parlour.* These are expensive to install, hence are better suited to very large milking herds.
- *Robotic milking.* This will not be discussed in this book.

If in-parlour feeding is required, it is best to equip milking parlours with grain feeders that allow each cow to be fed in ratio of her milk production. If cows expect to be fed while being milked, they will enter the parlour more readily, thus saving some labour. Each cow can be manually fed her ration using a measuring scoop or an electronically controlled concentrate dispenser. Alternatively, all cows can be fed the same quantity from an automatic concentrate dispenser. Cows can also be trained to enter the parlour willingly with no in-parlour feeding, and this requires very good cow handling and management expertise.

An entrance to a milking parlour that is straight, with no turns will ensure a smooth and convenient operation. Once trained, stock will walk readily into the parlour. A single step 1 cm high will help keep manure from being carried into the parlour. An exit leading to an uncrowded area will facilitate cow flow. It should be narrow (70 to 80 cm depending on cow size) to keep the cows from turning around.

Milking hygiene is generally very poorly managed in many countries, particularly those where farmers are not adequately rewarded or penalised for their milk quality. However, it will become increasingly important as the milk processors invest in the industry and when farmers consider value-adding with other dairy products, such as cheese, that require very good quality raw milk.

Milk room and cooler

Sanitation is the primary consideration in handling milk, whether from a hand or machine-milked cow. Adequate supplies of potable water are essential for cleaning the milking equipment immediately after use. Hot water (85°C) mixed with a chemical detergent is required for effective cleaning and cold water used for rinsing. Milk should be handled in a separate area that can easily be cleaned and is free of insects, birds, rodents and dust.

With machine milking, the vacuum pump and the engine that powers it should be housed in a separate room.

The milk room should be well ventilated with a concrete floor sloped 20 mm/m to a drain and with masonry walls having a smooth, water resistant surface that can be easily and thoroughly cleaned. On dairy farms of sufficient size, the milk is cooled by chilled water circulated between an evaporative water cooler and a milk cooler (plate heat exchange), through which the milk is passed until it is adequately cooled.

Where milk is transported in cans, cooling can be accomplished through a coil, which is immersed in the can. The larger-scale farmer, having a pipeline milking system, and milk collection by a tanker will require a refrigerated cooler and holding tank.

5.6.2 Electronic cow monitoring systems

In most modern large-scale dairy feedlots, each cow is provided with an electronic eartag which is an integral part of the farm's complete cow milking, feeding, breeding and veterinary recording system. Historical data on individual cows, together with graphical user interface, allow staff to monitor lactation curves, milk fat and protein contents. In addition, individual historical records of ailments, treatments and culling allow farm staff to evaluate their management practices.

The individual cow's milk yield can be automatically recorded. Sudden changes in daily milk yields are one of the fastest detectors of animal health problems in newly calved cows, with the first 3 or 4 weeks of lactation containing the majority of all health issues.

Cows can be grouped according to their milking time for more effective use of the parlour capacity as well as identifying, for isolation, those cows requiring special attention or treatment. Any animal requiring special attention can be separated via an automatic sorting gate.

As part of the electronic recording system, activity meters continuously collect and transmit activity data on each cow which provides valuable information, in addition to field observations, on cows' reproductive cycles hence the herd's breeding program. Low activity can also be monitored which may be due to disease. Following input of cow group feed intake data, feed costs and unit milk returns, computer printouts can provide business data on milk income less feed costs (MIFC) or Feeding Profits.

The rapid evolution of electronic monitoring equipment and computer systems will no doubt lead to many new objective assessments of cow productivity, herd performance and farm profits which, with further developments, will become more cost effective. One such example are mini-microphones attached to cow collars that can monitor rumination patterns throughout the day and night which provide valuable insights into individual animal reactions to changing heat stress and feeding management. An automatic accurate recording of cow respiration rates would be a very useful tool for the humid tropics.

5.7 Summarising considerations for shed designs

These elements have been summarised in Table 5.1 for concrete floored, feedpad sheds and in Table 5.2 specifically for free stall sheds. As previously mentioned, feedpads are concrete feeding areas for cows that also spend considerable time in between grazing pastures. The same equipment is required for feedpads as well as for any indoor stock feeding system, such as forage conservation, effluent handling and feed mixing and feeding out.

Various cowshed plans are presented in the Appendices, namely:

- Appendix 4; A dairy shed to hold 50 milking cows, 40 calves, 10 dry cows and 40 young replacement heifers or dairy beef animals and 3 herd bulls.
- Appendix 5; A free stall shed.
- Appendix 6; Cooling systems for a free stall shed.

Table 5.1. Elements to consider when designing feedpad sheds.

Elements	Considerations
Cow and feed alley surfaces (concrete)	Grooved (parallel to alley length) and broomed to provide maximum traction Provides good hoof support; flat surface with no irregularities Minimum hoof injury; no sharp edges or protruding aggregate Typical grooving is 12–25 mm deep, 20–25 mm wide and spaced at 100–125 mm intervals Edge of groove has no ridges Newer patterns have small grooves between larger grooves, to provide more traction Durable pad surface to withstand wear from feet, manure and water Rubber matting on turning area
Loafing area	Located adjacent to shed Run-off factored into effluent disposal design
Roof and wall type	Steel construction Gable style roof Removable wall panels, sliding doors or curtains to aid ventilation
Roof design	Lower roof pitch for slow air movement (1:4) Steeper roof pitch for faster air movement in warmer climates (1:3) Provide continuous open ridge to promote air movement, for convective heat dissipation Minimum eave height of 5 m at roof pitch 9 m In open side sheds, provide eave overhang of at least 1.5 m
Ventilation	Provide clean dry air to cows using passive or mechanical ventilation Passive ventilation is most economical Cross-ventilation provides air more readily to entire building
Machinery access	Minimum of 3.6 m clearance for entry and exit Minimum of 7 m for internal ridge height, for working machinery Minimum of 15–18 m turn around areas for tractors and machinery; check with manufacturers

Table 5.2. Elements to consider when designing free stall sheds.

Elements	Considerations
System design	Allow for easy cow movements around the shed Stalls should be comfortable, dry and clean Should allow for regular manure removal Provide easy access for machinery to feed out and clean Allow for multiple routes from stalls to feeding area, to overcome dominance within herd Feeding space and free stalls should be proportional Stalls, feeding and watering spaces should be arranged in modules
Modular design	No more than 60 stalls per module Modules can be manipulated to match site constraints
Stall sizing	Dependent on cow size (weight and body dimensions) Size based on largest 25% of cows in group Provide for adequate lying down as well as necessary forward and sideways lunging to stand
Stall slope	4% slope from front to rear 3% lateral slope, so cows lie in same direction facing down slope, to minimise teat injuries
Stall curb	Minimum of 200 mm above alleyway
Stall design	See text
Alley width	3–3.6 m wide avoiding sharp turns
Crossovers	Locate every 20–25 stalls, that is every 24–30 m 4.5–4.8 m wide, so 1 cow can drink and 2 cows pass behind Elevate floor slightly above stall curb, allow crossover to drain into alley
Bedding material	Inorganic (sand, rubber) Organic (sawdust, straw, hay, composted manure, rice hulls) At least 100 mm thickness of rice hulls, sawdust shavings or sand Consider hygiene, degree of compression when cow lies down, performance over time, how it will interact with effluent disposal system

6

Managing the human resources on tropical dairy farms

This chapter presents a template of how to access and manage the labour capacity required for effective management of tropical small and large-holder dairy farms.

6.1 Smallholder farms

The workforce on most smallholder farms consists mainly of the farmer himself and his wife's labour input, when available. Adolescent children may also be called upon to contribute some part-time labour. Depending on the herd size, the farmer may also employ additional workers for routine chores such as harvesting forages, feeding stock, milking cows or cleaning out cowsheds.

After considering the diversity of staff and their skills, employed on large-scale operations, it is apparent that in order to manage a productive and profitable dairy farm of any size, there are many skills required. Therefore, a successful smallholder farmer is a very competent person with many skills in both production technology and business.

6.2 Large-scale dairy farms

The size of the workforce on larger farms depends on the intensity of the production system, the herd size and how the farm sources its forage supplies. For example, with contract grown forages, less farm labour is required.

Large dairy production systems succeed or fail, generally, for two main reasons:

1. The ability of management to have appropriate goals and plans in place, make appropriate decisions, develop appropriate Standard Operating Procedures (SOPs), motivate their staff, as well as monitor operations well.
2. The capability and ability of staff to successfully implement the SOPs and be focused on the business being successful.

The initial procedures for the selection of management and operational staff, is immensely important in achieving success. Appropriate Position Descriptions (PDs), including levels of experience (and to a lesser extent qualifications), are critical to attract and retain the most appropriate staff. Following their appointments, induction and review programs need to be developed. All staff must be trained to carry out the SOPs developed for each role. In a new large-scale dairy development, it is especially important that this training be completed before the cows commence milking.

The logistics of human resource programs are then:

- develop PDs and reporting structures
  - PDs include job title and summary, responsibilities, expectations and physical requirements for the job
- advertise positions, interview and select staff on merit
- develop staff induction and review procedures
- develop staff Standard Operating Procedures (SOPs)
- train staff before heifers calve down and enter the milking herd
- upgrade staff numbers as cow numbers increase
- set goals and establish standards
- establish expectations and clearly communicate them to staff
- recognise and reward good employment outcomes
- weed out poor performers
- treat employees fairly and consistently
- seek feedback from employees.

The following is a detailed list of staff positions (and their PDs) required to fully staff a very large-scale dairy farm, with say 1000+ milking cows. It is assumed that this farm is owned by a collection of investors and has a management board to provide guidelines for corporate management as well as strategic goals for farm management decisions. The day-to-day farm management decisions would normally be left to the Farm Executive, which generally consists of senior farm staff, with or without guidance from specialist consultants.

1. **Farm Manager**; Reports to Farm Executive:
   - 'hands-on' manager of people, budgets and farm image
   - previous large-scale tropical or arid dairy management experience essential

- dairy nutritional experience desirable
- university degree level desirable but not essential
- finalises the annual operating plan developed by the farm executive
- Farm Executive to consist of Farm Manager (usually the chairperson), Veterinarian/Herd Manager, Feed Manager and Forage Production Manager
- responsible for overseeing the Veterinarian/Herd Manager; Feed Manager and Forage Production Manager and the Administration Assistant (see Figure 6.1)
- directly responsible for managing the milking staff.

2. **Administration Assistant**; Reports to Farm Manager:
   - must have previous administrative role experience
   - office management training preferable
   - must be very computer literate
   - liaises with all stakeholders and manages farm financials
   - inputs business and herd data into farm database and farm management system
   - depending on herd size, monthly cash throughput and total staff numbers, there may be more than one office-based administrative assistant
   - at least one of these staff should have accounting and business experience to work with the Farm Executive in preparation of annual budgeting and business review plans.

Figure 6.1: All farm staff must realise that because milk is over 85% water, cows must have a lot of drinking water every day, up to 150 L/cow/day. On many smallholder farms the cows are rarely provided with sufficient clean water. This tethered cow in Indonesia is only offered drinking water in the form of a concentrate slurry, hardly enough.

3. **Veterinarian/Herd Manager;** Reports to Farm Manager:
   - previous large-scale tropical dairy management and dairy herd data management experience essential
   - formal veterinary training, preferably with dairy nutrition experience
   - responsible for developing and implementing
     - ➤ the Animal Health Plan (calves, heifers, cows, bulls)
     - ➤ a Herd Health Data Management System
     - ➤ a Genetic Improvement Program
     - ➤ an Animal Health Program (such as lameness and mastitis)
     - ➤ other programs such as for Cow Comfort, Milk Production, Reproduction (Oestrous Detection, Artificial Insemination and Pregnancy Diagnosis, Calving and Disease Management)
   - will assist in developing the Feed Management Plan
   - will oversee the animal management staff
   - responsible for the management of Assistant Herd Manager, Calf/Heifer Rearer and general farm staff.

4. **Assistant Herd Manager;** Reports to Veterinarian/Herd Manager:
   - competent stockman with previous dairy cow handling/husbandry experience essential
   - certificate/diploma level animal husbandry training desirable
   - carries out the adult cow shed management, moving cattle (for milking and other), oestrous detection and recording, AI, oversees the cleaning of stalls, managing bedding and managing floor cleaning
   - assists Veterinarian/Herd Manager as required.

5. **Feed Manager**; Reports to Farm Manager:
   - previous large-scale dairy nutrition and herd feed management experience essential
   - certificate/diploma level animal dairy herd feeding/nutrition training preferred
   - responsible for developing feed management plan, with assistance from Herd Manager/Veterinarian and Nutrition Consultant
   - responsible for feed budgeting (1–2 year plan), feed ordering, ration formulation, ingredient and feed testing
   - responsible for overseeing ingredient delivery, feed storage, feed mixing, feed delivery, feed push up, feed clean up, feed recording
   - oversees Assistant Feed Manager.

6. **Assistant Feed Manager;** Reports to Feed Manager:
   - previous dairy herd feeding experience necessary
   - formal qualifications not required

- responsible for assisting feed storage, feed mixing, feed delivery, feed push up, feed clean up, feed recording
- assists Forage Production Manager when necessary.

7. **Calf/Heifer Rearer**; Reports to Veterinarian/Herd manager:
   - previous dairy cow, calf, heifer handling/husbandry experience essential
   - certificate/diploma level animal husbandry training desirable but not essential
   - responsible for the feeding and management of calves and heifers, including health and growth monitoring (incl. weighing) and reporting (see Figure 6.2)
   - assists Veterinarian/Herd Manager as required.
8. **Forage Production Manager**; Reports to Farm Manager:
   - previous forage production (for dairy herd) and tractor driving experience essential

Figure 6.2: Milk-fed calves are commonly fed their milk from a teat, or on many smallholder farms, direct from the cow. Calves drink water from a bucket, so why not milk as well? These calves in a Vietnamese dairy feedlot are fed calf milk replacer, which is cheaper than fresh milk, direct from a bucket.

- certificate/diploma agronomy level preferred
- experience in growing and conserving forages (and other feedstuffs) for dairy cattle essential
- responsible for planning, planting, tillage, irrigating, management, harvesting, and ensiling/conserving forage for farm
- responsible for overseeing Forage Production Assistant.

9. **Forage Production Assistant**; Reports to Forage Production Manager:
   - previous dairy cattle and tractor driving experience essential
   - no formal qualifications necessary
   - responsible for assisting in planting, tillage, irrigating, harvesting, ensiling forage for farm
   - when necessary, to assist in effluent management, feeding, shed cleaning and animal husbandry duties etc.

10. **Milkers**; Report to Farm Manager:
    - no formal qualifications necessary
    - preferably with veterinary livestock assistant training
    - responsible for the effective milking and milking hygiene management, milking and parlour cleaning and herd recording data collection.

11. **Senior Stockman**; Reports to Veterinarian/Herd Manager:
    - certificate/diploma level animal husbandry training desirable but not essential
    - responsible for routine animal husbandry tasks, including dehorning/disbudding
    - manages all foot trimming as well as other routine tasks as determined by Veterinarian/ Herd Manager.

12. **Farm Hands/Roustabouts**; Report to Veterinarian/Herd Manager:
    - previous tractor and dairy cow handling/husbandry experience essential
    - no formal qualifications necessary
    - full-time positions, used on an as-needs basis
    - manage effluent disposal
    - assist in calf rearing, heifer management, cow management, bull management, barn cleaning, bedding management
    - also assist in forage production and feeding
    - often responsible for gardens and physical appearance around farm
    - could also back up milker roles.

13. **Maintenance Manager**; Reports to Farm Manager:
    - previous farm machinery, equipment, building and animal handling equipment maintenance experience essential
    - certificate/diploma level machinery maintenance training desirable but not essential

- responsible for maintaining all farm machinery, equipment, buildings and cattle handling equipment
- oversees upkeep of the grounds and gardens
- role could also cover some security responsibility.

14. **Milk Chemist**; for milk analyses:
    - full or part-time, depending on milk throughput and value-adding.
15. **Security staff**; Night (and day) shift security staff. They could also be responsible for biosecurity on vehicles and staff entering the farm complex.
16. **Expert Consultants;** report to Farm Manager:
    - employed on an 'as needs basis'
    - dairy nutritionist; to advise on ration formulation
    - veterinarian; to advise farm veterinary staff on specific problems as they arise
    - agronomist; to advise on forage production
    - irrigation expert; to advise on crop irrigation (effluent disposal and irrigation water)
    - silage expert; to advise on effective silage making and storage strategies
    - cow cooling expert; to advise on establishing and operating cow cooling equipment
    - milk processing specialist; to advise on the establishment and operation of value-adding programs, if considered as part of the business plan
    - accountant and business consultant; to provide the necessary expertise (if unavailable among office staff) for achieving future farm development.

7

Best management practices for smallholder dairy farmers

This chapter presents an overview of the essential on-farm practices necessary to effectively run a smallholder tropical dairy to sustainably produce a consistently saleable product.

Best Management Practice (BMP) and Quality Assurance (QA) are processes for describing and implementing the most suitable procedures for a particular set of tasks to achieve a desirable outcome. With something as complex as running a dairy farm, it is best to partition the major outcome, that is profitable milk production, into several sets of management decisions that producers must make such as those summarised in Figure 3.1 in Chapter 3. These include growing productive forages, efficient feeding management, hygienic milk harvesting, effective animal health and rearing replacement heifers. Essentially BMP means:

- saying what you should do
- doing what you say
- recording what you have done

and hence improving the activities of your dairy operation.

QA programs are in their infancy on Asian dairy farms. With increasing focus on meeting customer requirements, QA programs are likely to become an integral part of dairy enterprises in future years and should focus on people, environmental and animal welfare, as well as product quality and safety management.

7.1 The importance of record keeping on dairy farms

Farm records are an integral part of good dairy farming practice. They help farmers in many ways such as making:

- herd and farm management decisions
- financial accounting and decision-making
- identifying problems
- planning for the future
- determining whether farm production targets have been met.

The major criteria for good record keeping are:

- *They must be useful.* Unless data are used in some future time and turned into information in making management decisions, the information should not be recorded at all.
- *They must be kept in such a form that they can be easily converted into information.* Before keeping a record, the eventual end use must be decided upon so that the form in which the data are recorded will facilitate later analysis and interpretation. Too often the end use is not considered and then usefulness is questionable.
- *The system must be simple.* Farming is a busy enough profession without burdening farmers with a complex record keeping system.
- *Duplication must be avoided as much as possible.* Some data may have to be recorded more than once in different forms and this can be done at a later date.
- *They must lead to actions being taken in the near future.* Information must lead to quick actions.

In other words, they must be easy to:

- record
- regularly update
- understand
- access
- summarise into information.

This discussion will not include financial and staff working times, but be limited to farm production and animal health information.

The keeping of farm production records can be divided into the five main activities, namely:

- animal identification
- herd breeding records
- individual cow milk production records

- health records
- feeding records.

7.1.1 Animal identification

Information should be recorded when an event occurs. Putting this off leads to omissions and errors. A good herd recording system will include identification of each calf as soon as possible after birth. In addition to the identification name or number, the calf's birth date, its size, name and its sire and dam's names should be recorded on a simple identification record.

Such information is useful to dairy farmers when:

- a sire is to be selected and the farmer wishes to avoid in-breeding
- heifers are evaluated for breeding and superior bulls must be chosen
- evaluating overall herd reproduction and determining the age of heifers at first heat
- knowing the age at which a heifer should be targeted for breeding
- determining whether an animal is an appropriate size for its age
- comparing genetic lines across an area (with other dairy farmers) to determine which animals to cull
- determining which animal should be culled on the basis of age.

7.1.2 Herd breeding records

A farmer with records based on identified cows can improve his breeding management by being able to determine such matters as: the date at which to dry a cow off; knowing when a cow should deliver a calf; highlighting poor insemination or bull services; establishing breeding dates and feeding programs; identifying calf, sire and dams; and determining the date for pregnancy testing. Breeding records must be up-to-date and easily accessible. They should be kept on a computer or in a dry and clean place within the dairy shed in a position which the farmer passes each day and can easily stop to update records.

Herd breeding records should include:

- the identification of an animal including its birth date
- name of sire and dam
- heat dates and comments
- calving dates and comments
- earliest breeding date
- service information
- pregnancy examination
- expected calving date
- drying off date
- any additional remarks.

7.1.3 Milk production records

Milk production records help the farmer determine which animals are bringing down the average yield and therefore income from the herd. Good milk production records can assist to raise milk production from an individual cow and a herd through specific management for individual animals. The more frequently individual daily milk production (and milk composition) data are recorded, the better. Daily herd milk yields and milk composition/quality data should be easy to access and record. Seasonality of milk production needs to be quantified to develop the most appropriate feeding strategies to address troughs in monthly milk yields.

7.1.4 Health records

Calling in a veterinarian to examine animals once they show a health problem is a more common approach than organising regular veterinary visits. With the use of records, veterinarians can gain additional information about the probable causes of ill health or poor reproduction in an individual animal to compensate for lack of familiarity with an individual arrival. Information concerning milk yields, quality of milk, dates of last change of status in reproductive and lactation terms, and other general information are of value in addition to the farmer's observations. Health issues and treatment information can also be recorded for individual stock on-farm. As well as recording veterinarian visits (date, purpose, diagnosis and treatment), vaccination details (type and date) and calf and cow mortalities, records are useful information in disease management.

7.1.5 Feeding records

These can vary from simple records of when feed was purchased and for which category of stock, through to details of changes in feeding programs (such as feed types and daily quantities fed per cow). The more detailed the feeding records, the better the ability to assess feeding efficiency (milk produced per unit feed consumed) on the farm.

The best way of recording such information is using a daily diary kept in the office where changes in herd and feeding management can be routinely included. This diary could also be used to document appointments or other planned important farm activities.

An additional aid to farm record keeping is a small pocket-sized notebook carried by the farmer (and staff) in which they record any unusual observations, changes in routine management or other noteworthy farm events. Computers are becoming more common to aid record keeping, collation and interpretation although a simple notebook and hand calculator is better than relying on memory.

7.1.6 Action lists

This term refers to details of particular stock requiring particular attention.

- ***Daily action lists*** would include cows likely to show heat, cows due for drying off, through to cows and heifers due to calve.
- ***Weekly action lists*** could include drenching of young stock, changes in cow and heifer feeding groups, through to heifers now ready for insemination.
- ***Monthly action lists*** would include routine veterinary examination, such as cows and heifers for pregnancy testing. It would also include cows that have aborted within the last month, re-examination of cows previously treated by the veterinarian within the last month, cows not observed to be on heat after 42 days from calving and also cows still not pregnant after 110 days post-calving. Other stock that should be examined would be cows with irregular cycles, cows confirmed in calf but have cycled again and any other problems listed in the daily diary. This would also include vaccination and parasite control requirements.

Examples of farm record pro formas are presented in Tables 7.1 to 7.5.

Table 7.1. Cow identification record within the herd.

Cow number						
Breed						
Date of birth						
Sire						
Dam						
Calf 1						
Date of birth						
Bull/heifer						
Sire						
Calf 2						
Date of birth						
Bull/heifer						
Sire						
Calf 3						
Date of birth						
Bull/heifer						
Sire						
Calf 4						
Date of birth						
Bull/heifer						
Sire						

Table 7.2. Fortnightly milk production record card for each cow (L/cow/day).

Cow identification:							
Day	**Mon**	**Tue**	**Wed**	**Thu**	**Fri**	**Sat**	**Sun**
Week:							
AM yield							
PM yield							
Day							
Week:							
AM yield							
PM yield							
Day							

Table 7.3. Breeding record card for each cow at each mating.

Cow identification: .... **Lactation number: ....** **Date of last calving: ....**				
Date on heat				
Service date				
Bull/AI				
Sire breed				
Pregnancy check Date and result				
Date to dry off				
Expected calving date				

Table 7.4. Vaccination record for the herd.

Date							
Routine vaccination record							
A*							
B							
C							
D							

* A, B, C, D indicate type of vaccination.

Table 7.5. Veterinary treatment record for the herd.

Date	**Animal number (s)**	**Clinical symptoms**	**Diagnosis**	**Treatment**	**Outcome**

The remainder of this chapter lists a range of BMPs for the day-to-day feeding and herd management of smallholder dairy (SHD) farmers in the tropics. It is split into two types of BMPs, first forage production and second, managing the dairy herd. The first is discussed under the banner 'Producing quality forages on smallholder dairy farms' while the second has been grouped under the banner 'Golden Rules for tropical smallholder dairy farming'.

This chapter simply lists the BMP, not detailing the scientific logic behind each one. There are so many BMPs (158 of them) with virtually every one of them discussed in detail in one or several of the senior author's previous five books on tropical dairy farming. Another section of this chapter discusses Key Performance Indicators (172 of them) and as with the BMPs, these are simply listed, rather than quantified (except for those listed in Table 7.6).

7.2 Producing quality forages on smallholder dairy farms

There are four basic principles of producing quality forages. These are:

1. Select the most appropriate forage species for the region.
2. Prepare the forage production area for sowing.
3. Manage the crop, particularly with adequate fertiliser to optimise growth and quality.
4. Harvest the crop at the best stage of maturity for nutritive value.

7.2.1 Selection of forage species

Key questions to ask when selecting varieties of forages to grow:

- Are there advantages over local varieties?
- Are they adapted to the climate and soils?
- Do they suit the intended use?
- Do they fit into the particular farming system?

More specific questions to ask are:

- Have they been tried and found successful in the region?
- Do they suit local farming systems and ecological conditions?
- What extra inputs are required, such as seed costs, labour and fertilisers?
- Will their extra cost return a profit?
- What are the risks of crop failure?
- Do the seeds come from a reliable source of supply?

No forages grow well everywhere. Some grow well in acidic soils while others do not. Some grow well in cooler areas, while others do not. Forages can survive in areas where they are not adapted but they may not thrive. It is important to choose forages that are adapted to local soils and climate.

7.2.2 Important soil and climatic factors

Important climatic factors affecting forage adaptation are the length of the growing season, air temperatures, soil fertility, soil pH and drainage.

The key properties of a soil that influence forage selection and plant growth are:

- soil texture: light (sandy, sandy loam), medium (loam, clay loam), heavy (heavy clay)
- soil fertility: low, medium, high. Soil fertility is the ability of the soil to hold and release nutrients for plant growth, rather than the amount of nutrients in the soil at any one time
- soil pH: strongly acid (5.0), acidic (5.0 to 6.5), neutral (6.5 to 7.5), alkaline (>7.5)
- drainage: well drained, moderately drained (occasionally waterlogged), poorly drained (frequently waterlogged)
- level of soil salinity: low, medium, high
- level of available Al (aluminium) and Mn (manganese): low, medium, high.

Other important key descriptors of forage plants are:

- latitude: tropics, subtropics
- altitude (m above sea level): 0–1000, 1000–2000, 2000–2500, >2,500
- annual rainfall: from 200 to 6000 mm/yr
- family: grass, legume, other (e.g. trees)
- life cycle: annual, perennial
- longevity of forage crop: long-term (>4 yr), short-term (<4 yr)
- purpose: grazing, cut and carry, conservation, hedgerow or living fence
- defined dry season: <6 months, >6 months
- inundation: 1 wk, 1 month, >1 month
- stem habit: erect, prostrate, climbing, stoloniferous (stems growing along the ground), rhizomatous (root nodules that convert atmospheric nitrogen into plant nitrogen)
- shade environment: moderate (30–50% shade), dense (>50% shade)
- frost intensity: light, heavy.

7.2.3 Agronomy of forage crops

The key issues to ensure high yields of quality forage from any forage crop are:

- Preparation of the soil for sowing, to ensure a good establishment of the crop and minimum weed invasion.
- Fertilising the crop with sufficient soil nutrients, using inorganic fertilisers to supplement dairy shed effluent. Ensure fertilisers are routinely applied after every (or at least every second) forage harvest.
- Having a crop harvesting program to ensure quality forage as well as adequate forage yields.

- Appreciating the difficulty of recommending specific harvesting regimes for all crops in every situation, because the rate of forage growth depends on variety, rainfall and irrigation, soil fertility and harvest interval.
- Wilting the harvested forage is one way to improve forage intakes through reduced plant water loads.

7.2.4 Ensuring year round forage supplies

The seasonality of rainfall and soil temperature does not allow for regular growth of forage crops throughout the year. High growth rates during the wet season provide the opportunity to harvest excess forage for conserving and storage to be fed out during the dry season when forage sources are usually in short supply.

Excess forages can be conserved as hay or silage. However, good ensiling practices generally produce better quality roughages than do hay because less time is required to wilt the feed, during which time the forage loses nutrients, causing a reduction in feed quality. Hay making requires a longer period of rain-free days, which are often rare in the tropics during the wet season when feed excesses generally occur.

7.3 The Golden Rules for tropical smallholder dairy farming

The following Golden Rules arose out of an FAO program entitled 'Smallholder Dairy Development Program (SDDP)' that operated in Thailand, Bangladesh and Myanmar from 2011 to 2015 (Asia Dairy Network 2015). A recent innovation was to crystallise farm management practices into eight Golden Rules that covered the entire range of on-farm herd management practices, from rearing young stock and feeding milking cows through to reproductive, animal health, housing and heat stress management, culminating in milking hygiene. For each Golden Rule, the SDDP developed a glossy brochure highlighting the key seven to eight BMPs to provide a simple set of 'take home messages' that could easily be translated into the language of the target country. To date, they are freely available in English and the languages of the three SDDP participating countries. We have also translated them into the Indonesian and Malaysian languages.

Forage production was not included as one of the Golden Rules because of the diversity of soil and climatic variables in the tropics and the wide variation in BMPs that can produce quality forages.

The 158 BMPs for the Golden Rules have been grouped into two types. First, for each Golden Rule, there are the seven or eight BMPs that are included as part of the original Golden Rules brochure. These are the simple 'take home messages' associated with hand drawings in the brochure and these have been numbered and highlighted in italics in the lists below. Second, there are the remainder of the

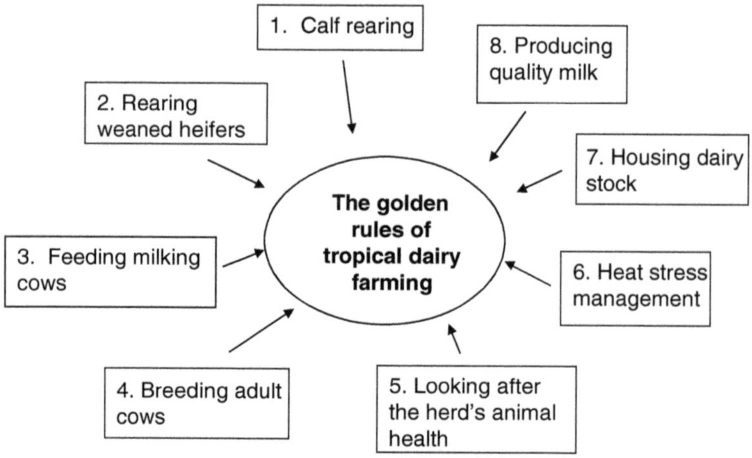

Figure 7.1: The eight Golden Rules of tropical smallholder dairy farming.

158 BMPs that are additional to the original 'take home messages' – these have not been numbered and they appear as non-italic bullet points in the following lists. The total number of BMPs for each Golden Rule are shown after each subheading.

7.3.1 Golden Rules of calf rearing (21)

1. *Calves must be born under hygienic conditions. Calving cows must be provided with a clean dry area in which to calve down.*
2. *The calves' umbilical cord should be sprayed with iodine solution (7%) immediately after birth.*
3. *Calves must be given a good drink of quality colostrum early in their lives. Each calf should receive 4 L of colostrum within 6 h of birth.*
4. *Calves must be always provided with adequate clean drinking water.*
5. *Calf pens should be provided with good natural ventilation against heat stress and also protection from any severe cold weather in winter.*
6. *Calves can be weaned off milk when each one consumes ~1 kg/day of calf concentrate. With the correct feeding management, this should occur by 3 months of age.*
7. *Permanent identification of each calf (with an eartag) and good record keeping will always benefit the farmer in the long run.*

- The colostrum must be of good quality with adequate amounts of antibodies; this can be tested by simple and cheap colostrometers or refractometers in a laboratory but thick creamier-looking colostrum is likely to have more antibodies than thin, diluted-looking colostrum.

- The colostrum can either be given by assisting calves to drink from their dam, or fed from a teat bottle or via stomach tube.
- Milk-fed calves must be fed a balanced diet of milk, concentrates and a small amount of roughage from the second week onwards. This varies with the age of the calf. The concentrate stimulates development of the rumen wall while the roughage stretches the rumen wall.
- Calf milk replacer is generally cheaper and can be as good quality as whole milk. Ask your local veterinarian or dairy cooperative for advice.
- There is no benefit whatsoever in allowing calves to suckle their dams. Milking cows do not require a suckling calf to stimulate milk letdown, and this can even lead to poor letdown behaviour over time.
- Calves can grow just as well if fed their milk once each day rather than twice each day as the clotted milk is slowly digested in the calf's stomach over the next 10 to 12 hr.
- Do not feed milking cow concentrate to young calves as it does not contain enough protein for a rapidly growing calf. Find a special calf-rearing concentrate.
- Calf-rearing concentrate should include minerals and vitamin supplements to provide all the essential nutrients for heifer growth.
- Calves should be reared in clean, manure-free and dry pens, ideally with no more than four calves in a pen.
- Minimise stresses by following a set feeding routine each day, especially with reference to time of day, milk volume and temperature.
- Scouring calves should be given electrolyte fluid replacer and antibiotics only under veterinary advice.
- Develop a disease action plan that includes good hygiene, isolation of sick calves, and TLC (tender loving care). Drugs should only be used as a last resort to complement a well-managed system.
- If more than 10% of your milk-fed calves die within the first three months of feeding (that is 1 out of every 10 calves), there are problems with their feeding and herd management. Ideally such death rates should be less than 5% (that is 1 in every 20 calves).
- A good calf rearer knows which calves are likely to be sick tomorrow.

7.3.2 Golden Rules for rearing weaned heifers (13)

1. *Weaned heifers require a balanced diet of quality roughages and concentrates. This varies with the age of the calf.*
2. *Their rumen is still too small for them to only be fed roughages so also feed quality concentrates, containing at least 16 to 18% protein. This could simply be good quality milking cow concentrate.*

3. *All dairy animals require regular access to clean drinking water.*
4. *Well-reared heifers will show signs of heat within the first 10 to 12 months of their life.*
5. *Well-reared heifers have well-developed bones, hence a good frame size, as assessed by wither height.*
6. *Well-reared heifers will calve down at 24 to 27 months of age, not 30 or 36 months of age, as is generally the case in most Asian countries.*
7. *During their lifetime, well-reared heifers will produce sufficient extra milk and calves to not only pay for their higher feeding costs but also yield substantial profits.*

- Their concentrate should include minerals and vitamin supplements to provide all the essential nutrients for heifer growth.
- If allowed to lose weight or grow very slowly for lengthy periods, heifers will not achieve their potential bone frame size and that can lead to problems in later life.
- Low mating live weights can lead to calving difficulties 9 months later. This can reduce milk yield and increase the number of days to the second conception.
- Poorly reared heifers are stunted and often have a big 'pot belly'.
- If more than 30% of your young stock dies before they calve down (that is 3 out of every 10 calves and heifers), there are problems with their feeding and herd management. Ideally such death rates should be less than 10% (that is 1 in every 10 calves and heifers).
- Internal parasite control is necessary if heifers are being grazed or fed green feed from manure-fertilised forage or pasture.

7.3.3 Golden Rules for feeding milking cows (23)

1. *Milking cows have high requirements for water, which should be supplied separately as clean drinking water rather than as part of any concentrate slurry.*
2. *Feed sufficient quality forages (20 to 40 kg fresh forage/cow/day). The daily amount will depend on the cow's live weight and milk yield and the farmer's available forage resources.*
3. *Supplement feed with concentrates that are formulated to overcome specific nutrient deficiencies.*
4. *Consider wilting the forage, by leaving it out in the sun during the day before chopping it up, to reduce its moisture content and encourage the cows to eat more of it, hence produce more milk.*
5. *At any one time, 60% of milking cows at rest should be ruminating. This is a good reflection of the overall good herd management, which includes appropriate feeding management.*
6. *If concerned about unbalanced diets in the milking herd, closely monitor the manure characteristics, changes in feed intakes, changes in milk yield and*

composition (fat and protein or solids-not-fat) and the proportion of cows that are ruminating.

7. *Remember that potentially higher yielding cows are more susceptible to other farm constraints such as insufficient quality feed, heat stress, poor animal health and the limited management skills of the farmer.*
8. *Farmers need to develop the skills to identify when cows are on heat. This requires consistent observations, including night-time observations.*

- Milking cows may cost more to feed better but they will do more than simply return this investment and will also yield substantial profits in producing more milk and more calves over their lifetime.
- There is no benefit whatsoever in providing concentrates in the form of a slurry.
- When planning milking cow feeding programs, it is important to feed more concentrates to higher yielding cows.
- It is better to feed green forages rather than dried ones, such as rice straw, as long as cows are consuming enough dry matter.
- Selection of forages to grow for livestock fodder should be based on those most suited to the soils, climate and local farmer skills.
- Growing forages requires additional inorganic fertilisers as well as cow manure and shed effluent.
- You must feed the grass as well as feed the cows. Use inorganic fertilisers for forage production.
- The optimum time to harvest forages should be based on their nutritive value rather than their total forage yield. Over-mature forages have a low nutritive value.
- Concentrate supplements should be formulated to provide sufficient energy, protein, minerals and vitamins for good consistent milk yield and fertility.
- Calcium and phosphorus supplements are particularly important for milking cows.
- Sourcing ingredients for concentrate supplements should be based on their relative costs of feed energy and protein.
- Cows should be able to lie down and rest for as long as they need to (up to 11–12 hr/day), on comfortable and dry bedding as this increases salivation, rumination, blood flow to the udder and hence milk yield. It also reduces lameness.
- It is best to set realistic target milk yields based on the genetic quality of the cows and the feed (forages and concentrates) available.
- It is more efficient to feed fewer milking cows better but the ideal target milk yields should depend on the farmer's resources (for feed supplies), management skills and the motivation to have high yielding cows.
- Cows tend to moo when they are on heat or are hungry. Don't ignore their cry.

7.3.4 Golden Rules for breeding adult cows (17)

1. *Milking cows need to be well fed to rebreed. If cows are too thin or over fat, they will have poorer fertility.*
2. *The voluntary waiting period, or days between calving down and rebreeding, should be no more than 50 days.*
3. *Look for signs of heat 18 to 24 days after their last heat, if they are still non-pregnant.*
4. *The best conception rates occur on insemination (natural or artificial) 4 to 12 h after the first signs of heat.*
5. *It is important to keep good records of inseminations and natural matings to ensure good herd reproductive performance and to know when to seek veterinary advice.*
6. *It may be necessary to cull additional cows if more than 15% of the herd have more than 16 months between calvings.*
7. *Each month farmers should identify non-pregnant cows that have calved more than 80 days beforehand but have not been detected to be on heat. The veterinarian should be asked to examine their reproductive tracts for abnormalities.*

- Unhealthy cows, such as those with retained placentas, lameness, mastitis or infections in their reproductive tract, have poorer fertility.
- One of the biggest challenges when rebreeding cows is to feed them an adequate well balanced diet in early lactation (the first 100 days of their lactation). Without sufficient feed nutrients, the cows will use their body reserves to produce milk, quickly lose weight and not cycle (show signs of heat). Pay particular attention to first-calf heifers.
- Farmers need to develop the skills to identify when cows are on heat.
- Identifying which cows are on heat requires consistent observations, including night-time observations.
- High temperatures and humidities will markedly reduce the number of hours that cows can be successfully rebred, either by insemination or using a bull.
- With artificial insemination, it is important to use insemination technicians with reliable good inseminating skills and use semen that has been tested as good quality.
- Veterinary advice should be sought if too many cows show signs of ill health such as assisted calvings, retained foetal membranes, lameness, abortions or vaginal discharges more than 14 days after calving and other health problems.
- Cows with calving problems are more likely to have uterine infections that can last for weeks (even months) after calving down, even though the cow shows normal signs of heat and does not have any abnormal vaginal discharge.

- It is just not possible to get 100% conception rate (cows getting back in calf), no matter how many inseminations the cow has. Identify any suspected problem cows for veterinary inspection.
- But remember that poor feeding management during the dry period (in high yielding cows) and the first 100 days of lactation is the most common cause for repeat breedings. Heat stress will make it worse.

7.3.5 Golden Rules for looking after the herd's animal health (19)

1. *Develop a disease prevention plan that includes good hygiene, isolation of stock and a regular vaccination program.*
2. *Learn to quickly identify subclinically sick animals from their physical appearance and behaviour.*
3. *If an animal is obviously sick, then move it to an isolation pen in order not to infect other stock.*
4. *Lameness can be a major problem with cows living in sheds. As it can be due to trauma, unbalanced diets or overgrown hooves, its cause must be identified.*
5. *Providing cows with soft bedding, such as a dry soil or sand resting area or rubber mats improves cow comfort as well as reducing feet and leg problems.*
6. *A written herd health program should be developed that contains management actions for herd health to prevent and control diseases and disorders specific for the region.*
7. *Do not use old drugs that are past their use by date to treat sick animals.*

- Drugs should only be used as a last resort to complement a well-managed system.
- Keep the hospital pen clean and occasionally sterilised.
- Learn to monitor respiration rates to quickly identify cows that are sick and heat stressed, to take appropriate action and cool them down.
- Develop a regular vaccination program to guard against local disease problems.
- Develop a regular dosing program to guard against internal parasites, which can even be introduced with cut forages.
- Be aware of external parasites that can adversely affect cow performance, therefore, treatment must be anticipated for such infestations in the herd.
- If lameness is most likely due to trauma, ensure floors are not too abrasive and all stones and broken pieces of concrete or metal have been removed.
- Consider foot trimming if the claws appear overgrown, and provide footbaths which contain formalin (to harden up the hooves) or copper sulphate (to treat sore feet).
- Sore feet can be due to feeding problems (acidosis) and if identified, rumen buffers and additional forages should be fed.

- The herd health plan should contain actions for foot care, mastitis, parasite control, infectious diseases and calf care.
- Remember that care and support for a sick animal is just as important as the veterinarian developing a treatment protocol.
- If advised to do so, keep drugs in dark and cool place.

7.3.6 Golden Rules for heat stress management (17)

1. *Hot cows lose their appetites. Very thirsty cows lose their appetites.*
2. *Heat stress will reduce the time cows can cycle, that is show signs of oestrus.*
3. *Heat stress will reduce the cow's fertility and if severe, can actually kill the developing foetus inside the pregnant cow.*
4. *It is better to feed cows during the night, when it is likely to be cooler, as well as during the day.*
5. *Extra water is essential on hot days or during summer for the cows to lose any extra body heat.*
6. *Sheds with low roofs and walls (which obstruct the wind) will reduce natural ventilation.*
7. *Spraying water on hot cows, even with a hose for just a few minutes, followed by placing them next to a fan (or better natural ventilation), will cool the hot cow. The water should be sprayed over the cow's trunk, enough to run off the cow's hide.*

- Cows eating a lot of feed to produce more milk, have a large mass of feed in their rumen being broken down by rumen microbes and this generates a lot of internal body heat.
- Cows feel the heat a lot more than humans do, so don't simply use your own level of comfort as a guide to how the cows are feeling.
- Cows must have regular access to clean drinking water, at least three or four times (preferably more frequent) each day, when they should be allowed to drink as much and for as long as they want.
- It is not always expensive to improve on the natural ventilation of any cowshed.
- A cow's heat stress can easily be monitored through counting the number of breaths she takes, that is her respiration rate.
- If the number of breaths is more than 70 or 80 per minute, the cow is heat stressed and requires extra attention. If the cow is breathing at more than 100 per minute, she is very heat stressed and requires immediate attention.
- Increasing air movement, even just using simple household fans, will improve cow cooling rates.
- Ideally cows should be able to go outside at night-time where they can cool off more effectively.

- Having a yard to walk around in at night makes it easier for cows to identify other cows that are on heat.
- Heat-stressed cows that drool saliva while panting are losing a lot of minerals from the body, hence require additional minerals in the concentrate formulation.

7.3.7 Golden Rules for housing dairy stock (30)

1. *Cows need 10 to 12 hr of undisturbed rest each day, preferably in comfortable stalls.*
2. *With well-designed walkways and resting stalls, comfortable lying surfaces and non-slip walking surfaces, cows suffer less lameness, can behave more normally and suffer less stress, have higher daily milk yields and longer productive lives.*
3. *Cows tied up all day every day on concrete floors are not comfortable.*
4. *Cows with hindquarters covered in fresh and dry manure are not comfortable.*
5. *Cows forced to continually lie on dirty hard floors when resting are not comfortable.*
6. *Cows that have difficulty lying down and standing up because their tie rope is too short are not comfortable.*
7. *Providing a portion of the daily feed in an outside yard encourages the cows to move around in the open.*

- Cow comfort is a new term to describe how cows 'feel' and can respond to better feeding and farm management.
- If cows are not 'comfortable', they will have poor reproductive and general health issues, will not produce more milk and more calves no matter how well they are fed.
- Comfort describes the housing conditions, the cow's ability to occasionally 'stretch her legs', the degree of heat stress she suffers and the way the farmer treats and handles her.
- Cows need a soft bedding which is kept clean and non-slippery.
- The best stalls are those that cows voluntarily enter and leave (called free stalls).
- Hot cows lose their appetites. Very thirsty cows lose their appetites.
- Sheds with low roofs and walls (which obstruct the wind) will reduce natural ventilation.
- Milking cows are unlikely to feel the cold anywhere in South-East Asia.
- Calves will rarely feel the cold anywhere in South-East Asia as long as they are dry, out of the wind and have loose hay to nest in.
- A cow's heat stress can easily be monitored through counting the number of breaths she takes, that is her respiration rate.
- If the number of breaths is more than 70 or 80 per minute, the cow is heat stressed and requires extra attention. If the cow is breathing at more than 100 per minute, she is very heat stressed and requires immediate attention.
- The presence of cobwebs in the roof area is indicative of low air movement, hence poor ventilation.

- Cows must have regular access to clean drinking water, at least 3 or 4 times (preferably more frequent) each day, when they should be allowed to drink as much and for as long as they want.
- Cows that have respiration rates exceeding 80 breaths per minute for many hours each day are not comfortable.
- Cows need to be allowed to walk around and 'stretch their legs' every day or so. This regular exercise decreases leg-related problems, mastitis, bloat and calving difficulties.
- It is not necessarily expensive to make cows more comfortable, hence more productive.
- Cows should never be made to fear any of the farm workers through their rough handling.
- Happy cows make happy farmers.
- Look after your cows and they will look after you.
- Cows feel secure when they know they have plenty of feed, have an escape route and know the stockman. They feel insecure on slippery floors, when very lame and when in the presence of unpredictable personnel.
- The body language of the cow is the best management adviser. If you can read the cow, you know what to do.
- Cows will tell you what they want.
- A good dairy farmer is one who extends his farming skills to cover cow psychology, cow behaviour as well as cow production technology and farm business management. In essence, he should be able to put himself 'inside the cow's skin' to develop the ability to 'think like a cow'.

7.3.8 Golden Rules for producing quality milk (18)

1. *Poor milk quality reduces the shelf life of liquid milk (that is pasteurised or ultra high treated or UHT) and is becoming more important for smallholder dairy farmers.*
2. *The most important point to producing quality milk is hygiene, both while milking the cows and handling the milk from the cowshed to the milk collection centre.*
3. *Milking staff should have clean hands and clothes, have no obvious signs of sickness and preferably use gloves.*
4. *Milking utensils should be thoroughly cleaned with detergent, sanitised and hung upside down to drain and dry out before the next milking.*
5. *Each milking cow should be provided with a separate cloth to clean the udders and teats before milking. It should then be washed, sterilised and hung up to dry in the sun before using at the next milking.*
6. *Cows' teats should be sprayed or dipped in 7% iodine solution following milking. Your local veterinarian or dairy cooperative should be able to provide this.*

7. *The California Mastitis Test (commonly called CMT) should be undertaken on each milking cow once each month until the herd reaches a low level of mastitis. Then it can be undertaken every three months.*

- Milk quality refers to the bacterial contamination in the cow's udder or the milking equipment whereas milk composition refers to constituents such as fat, protein, calcium and solids-not-fat.
- Cows must be encouraged to stand up for at least 30 min after being milked to allow the teat canal to become sealed off. This is easily achieved by feeding each cow immediately following milking.
- Mastitis is one of the biggest problems for smallholder dairy farmers throughout South-East Asia.
- Mastitis can be clinical (visually obvious) or subclinical (only apparent following the CMT).
- For every clinical case, there are generally 20 to 30 subclinical cases of mastitis.
- Infected quarters can lose 25% of their milk yield and produce only poor quality milk.
- Cows with mastitis should be milked last in the herd.
- Identified cases of subclinical mastitis should be immediately treated with antibiotics; this is called lactation therapy in milking cows. The milk must be discarded during treatment and for the next week or so, which is during the withholding period after treatment.
- Every cow should be given a dose of slow release antibiotic in each quarter when she is dried off. This is called dry cow treatment.
- Some cows will become carriers of mastitis and never be cured, so once identified as repeat cases they should be culled from the milking herd.
- Clean straw and laneway conditions, as well as managing heat stress will assist greatly in preventing mastitis.

7.4 Key Performance Indicators

The concept of Key Performance Indicators (KPI) is not new. However, it is when applied to quantifying the performance of SHD farmers in the developing tropics. Expressed simply, KPIs are diagnostic tools allowing farmers to improve their farm productivity, resulting in better financial performance. Farmers can use them to identify weaknesses in, as well as set specific targets for their farms. They are more likely to try to improve their systems if they know by how much they are less productive compared to others.

Each KPI provides a valuable insight into the farm resources and management skills of individual smallholders. However, it is important to prioritise them based on:

- their relevance to the farmer's current stage of farm development
- the farmer's ability to interpret the data and use it in future decision-making
- the ease and accuracy of collecting the necessary raw data.

Many Asian dairy farmers intuitively think about farm costs and returns. However, greater use could be made of formats allowing them to be more aware of the relative importance of all their financial inputs in terms of cost of production (COP) per kg of milk produced on the farm.

Knowing their cost of production allows smallholder farmers to determine their profit margins and this is critical to operating a sustainable dairy enterprise. Farmers must do more and better planning if they are to achieve greater profits. Profits are not something they end up with at the end of the year. Rather, profits are something for which farmers must plan.

Table 7.6 presents a range of and quantifies a series of Key Performance Indicators (KPI) to help farmers diagnose the strengths and weaknesses in their dairy enterprise. Farmers should use these indicators initially to identify these weaknesses in, rather than set targets for, their farm. Farmers are more likely to try to improve their systems if they know they are less productive compared to others.

Table 7.6. Ten key measures of smallholder dairy farm performance.

Measure	Questions to ask
Feeding management	
1. Stocking capacity	Is the farm carrying too many cattle for the available forage supplies?
2. On-farm forage production	How much of the farm's annual forage requirements must be purchased?
3. Forage quality	Is the forage being harvested or purchased at its optimal quality for milking cows?
4. Concentrate feeding program	What is the quality of the concentrates being fed and how much is allocated per milking cow?
5. Total feed costs	Are the forages and concentrates costing too much per unit of feed energy or protein?
6. Milk income less feed costs	How does this compare with those of other farmers with good feeding management?
Herd management	
7. % productive cows	What is the % of adult cows actually milking? What is the proportion of milking cows in the entire dairy herd, expressed as a percentage?
8. Pattern of milk production	What is the peak milk yield of the herd and what is its lactation persistency (rate of decline from peak milk yield)?
9. Reproductive performance	How many days after calving do cows cycle? What is the submission rate and the conception rate to first insemination? What is the inter-calving interval?
10. Heifer management	What is the pre-weaning calf mortality and the wastage rate of heifers from birth to second lactation? What is their age and live weight at first calving?

Such an approach may encourage farmers to look more critically at their cost structures. Expressed simply, this is a diagnostic tool to help identify production weaknesses adversely affecting financial performance.

The following series of 10 questions should be asked on any farm, big or small. Because more than half of farm costs are feed related, the first six questions are directly related to feeding management. Even though the remaining four are more related to overall herd management, they are still very much feed-dependent. For some of the questions, specific indicators relevant to particular farming systems can be developed. However, for others, there is no single indicator that farmers can work towards because the most correct answer is the higher the better for some (such as on farm forage production or forage quality) or the lower the better for others (such as total feed costs or calf mortality and heifer wastage rates). These indicators should be presented as ranges rather than a single value emphasising the fact that they are only guidelines. These 10 key measures or symptoms of poor farm performance for which diagnoses should be considered are presented in Table 7.6.

Actual values for these KPIs have been discussed in detail by Moran (2009b). They could be summarised as shown in Table 7.7.

Table 7.7. Actual values for KPIs following Moran (2009b).

1. Stocking capacity	8 to 12 adult cows per ha forage production
2. On-farm forage production	The more the better
3. Forage quality	9.5 to 10.0 MJ/kg DM ME, 12 to 14% protein
4. Concentrate quality	11 to 12 MJ/kg DM of ME, 16 to 18% protein
5. Total feed costs	Not too high, below 60% total cost of milk production
6. Milk income less feed costs	The higher the better
7. % productive cows	60 to 74% of cows in milking herd of those that have calved
	40 to 48% of milking cows in the total dairy herd
8. Pattern of milk production	High peak milk yield, 25 to 30 kg/cow/day
	Low rate of decline from peak, >8% per month from peak
9. Reproductive performance	55 to 60% pregnant within 100 days of calving down
	13 to 15% still non-pregnant within 200 days of calving down
	65 to 70% '80 day submission rate' (that is cows which were inseminated within 80 days of calving down)
	50 to 60 day voluntary waiting period before insemination
	45 to 50% conception to first insemination
	1.8 to 2.0 inseminations per conception
	12 to 14 months inter-calving interval
10. Heifer management	4 to 6% calf mortality to weaning
	20 to 25% 'wastage rate' (stock deaths between when they were born and their second calving)
	250 to 300 kg live weight at mating
	400 to 450 kg live weight at first calving
	26 to 28 months age at first calving

7.4.1 Ease of collecting raw data

Each and every one of the KPIs presented in Table 7.6 provides a valuable insight into the farm resources and management skills of individual smallholders.

The ability of smallholder dairy farmers to collect the raw data would vary greatly with their management skills, level of education, support from service providers and, of most importance, their motivation to want to use the particular KPI in their farm decision-making.

Some of the easiest to collect are the raw data to calculate the proportion of productive cows (KPI 7) as most smallholder farmers know the number of milking cows, dry cows, heifers and calves in their herd. In a matter of minutes, the senior author collected such data from a dozen smallholders at a dairy farmer conference in Vietnam a few years back, which indicated that most of these particular farmers had acceptable values for percentage of milking cows in the milking herd (KPI 7a) and percentage of milking cows in the entire herd (KPI 7b). That is why they were invited to attend the conference. Another 'easy to collect' dataset is the stocking capacity (KPI 1), as most farmers know their forage production area. These same farmers had eight to nine total stock/ha and five to six adult cows/ha, indicating that they did not overstock their farms. As they were selected to attend the conference, one could assume that they were considered to be 'good' farmers.

It is not difficult to collect raw data on pattern on milk production (KPI 8) as many farmers record daily milk yields from individual cows. The level of concentrate feeding (KPI 4) is another dataset readily available while many farmers could estimate their daily forage-feeding program (as a guide to KPI 2). Likewise, some of the reproductive (KPI 9) and heifer data (KPI 10) can be easily calculated from raw data on dates of inseminations, calvings and ages at first calving, while a chest girth tape can easily provide estimates of live weights of different classes of stock. Inter-calving interval is also relatively easily calculated and is a good general measure of herd health, nutrition and reproductive performance.

The above diagnoses require the calculation of many KPIs to allow a value judgement to be made on business performance. Many of these indicators are simply common ratios or proportions, assessing some level of output in relation to some level of input. Others measure success simply with numbers or amounts, such as target forage quality or heifer live weight. Although they are valuable guides, there is no all-encompassing or perfect indicator of business success. All indicators must be viewed within the whole business, with each one contributing only a part.

It is possible to achieve high performance in a KPI that does not translate into business financial success. If a farmer whose farm has very poor quality soils and may not be able to grow as good a quality forage as he can purchase, at a good price, close by, it would be more profitable to let someone else grow the bulk of his forages.

Low performance measures in some key factors, well below these KPIs, often lead to high performance measures in other key factors that can produce a false sense of security about the ability to achieve some of the production targets. One example is low peak milk yield and short inter-calving intervals in cows of low genetic merit. Because such cows are not 'genetically programmed' to use their body reserves to supplement the limited intakes of feed nutrients during early lactation, their live weight will hardly change and they may cycle soon after calving. If the farmer plans to improve the genetic merit of the cows by using imported cows or high-grade semen without improving the feeding management during early lactation, peak milk yields may not greatly improve while herd fertility is likely to drastically fall.

The list on Table 7.6 is an initial attempt to prioritise these indicators to develop a structured approach to addressing poor farm profitability. It must be stressed that no single KPI should be used in isolation to assess farm performance and hence profitability, as each one is the end result of interactions between many farm inputs. It is important to ensure there is a balance between their utilisation so that one production target is not achieved at the expense of others within the farming system.

The above are just a few of the KPIs available to assess a farm's performance. A more thorough collection of 172 KPIs related to the eight Golden Rule is presented below.

7.5 Selected KPIs quantifying each Golden Rule

1. Calf rearing

1. Average birthweight of female calf (kg).
2. Average weaning weight of female calf (kg).
3. Average weaning age of female calf (days).
4. Average daily growth of female calf (1–6 month) (kg/d).
5. % calves with diarrhoea.
6. % calves requiring veterinary attention.
7. % pre-weaning calf mortality.
8. % pre-weaning calf morbidity.
9. Clean dry area for cows to calve down (yes/no).
10. Iodine spray for umbilical cord (yes/no).
11. Are all newborn calves fed colostrum? (yes/no).
12. Calves milk fed from teat or bucket not from suckling (yes/no).
13. Permanently eartagged and recorded at birth (yes/no).
14. Have male calves been castrated before three months of age? (yes/no).

2. Rearing weaned heifers

1. Average age of heifer at first insemination (months).
2. Average weight of heifer at first insemination (kg).
3. Average weight of heifer at first calving (kg).
4. Average age at first calving (months).
5. % wastage rate (loss from the herd) from birth to second lactation.
6. Regular access to clean drinking water (yes/no).
7. Provided with quality ration of concentrates and limited forage (yes/no).
8. Crude protein content of heifer ration (%).
9. Wither height at say 12 and 24 months of age (cm).

3. Feeding milking cows

1. Good quality pasture fed per milking cow (kg/d).
2. Availability of good quality forage for cows feeding (months/yr).
3. Additional preserved forage fed per milking cow (kg/d).
4. Total concentrates fed per milking cow (kg/d).
5. Crude protein percentage of milking cow ration.
6. Frequency of offering drinking water to milking cows (times/d).
7. Source of drinking water and its quality.
8. Mean body condition score (/5 points); % of very lean cows and % very fat cows at 3 months of lactation.
9. Rumen score (/5 points); % cows with deeply hollow rumens.
10. Manure score (/5 points); % cows with any coarse particles in their dung and consistency of stiff balls (horse manure).
11. Number of stock per drinker or per cm of drinking trough, water flow (L/minute) and cleanliness of drinkers.
12. Stocking capacity (number of adult cows per ha forage production area).
13. On-farm forage production. How much of the farm's annual forage requirements must be purchased?
14. Forage quality. Is the forage being harvested or purchased at its optimal quality for milking cows?
15. Concentrate feeding program. What is the quality of the concentrates being fed?
16. Total feed costs.
17. Milk income less feed costs. Are the forages and concentrates costing too much per MJ of energy or kg of protein?
18. Proportion of adult cows actually milking (%).
19. Proportion of milking (and dry) cows in entire dairy herd (%).
20. Herd structure: replacement heifers (%), cows older than 4, 6 or 8 years (%).

21. Milking cow herd structure: Pregnant/wet (P/W), P/D, NP/W, NP/D (%); wet means lactating.
22. Peak milk yield of the herd (kg/d).
23. Pattern of milk production (e.g. rapid decline from peak).
24. Lactation persistency (rate of decline from peak milk yield) (%/month).
25. Average milk yield (kg/d).
26. Average days in milk (d).
27. Average lactation length (d).
28. Average production/lactation and production/cow/year.
29. Milk composition (%fat, %SNF, %TS).
30. On-farm forage production: Proportion of the farm's annual forage requirements that are purchased (%) or harvested off-farm (%).
31. Forage quality: Is the forage harvested or purchased at its optimal quality for milking cows? (yes/no).
32. Frequency of concentrate feeding of milking cows each day (times per day).

4. Breeding adult cows

1. Time from calving to first heat (days).
2. Voluntary waiting period (days).
3. Time from calving to first service (days).
4. Time from first service to conception (days).
5. Time from calving to conception (days).
6. Average services per conception.
7. Cows conceiving to first service (%).
8. Cows conceiving to artificial insemination (%).
9. Average inter-calving interval (months).
10. Average gestation length (days).
11. Cows aborted (%).
12. Cows requiring assisted calving (%).
13. Milking cows with breeding performance records (%).
14. Cows applied upgrading breeding program (%).
15. Native cows in the milking herd (%).
16. Purebred/high grade cows in the milking herd (%).
17. Assisted calvings; % cows calving/year.
18. Herd's submission rate.
19. Herd's conception rate.
20. Herd's 100 day-in-calf rate.
21. Herd's 200 day-not-in-calf rate.

5. Looking after the herd's animal health

1. Cows/heifers with health recording (%).
2. Cows/heifers received FMD vaccination (%).
3. Cows/heifers received internal and external parasite prevention (%).
4. Cows/heifers free of brucellosis (%).
5. Cows/heifers free of TB (%).
6. Cows/heifers free of Para-TB (%).
7. Cows free of mastitis (%).
8. Cows/heifers with health problems (%).
9. Culled cows/heifers with health problems (%).
10. Locomotion score (/5 points), % cows moderately to severely lame.
11. Clinical lameness cases per month.
12. Hoof score (/3 points), % cows with severe hoof inflammation.
13. Teat score (/4 points), % cows with rough callous ring around the teat ends.
14. Mastitis, clinical cases; %/month.
15. Mastitis, subclinical cases; %/month.
16. Mean number of coughs per cow per day.
17. % on-farm mortality.
18. % downer cows.
19. Sudden deaths/casualties; %/year.
20. Dull/obviously sick cows; % cows/month.
21. % cows with nasal discharge.
22. % cows with discharge from their eyes.
23. % cows with hampered respiration.
24. % cows with diarrhoea.
25. % cows with discharge from the vulva.
26. % dystocia in 12 months.
27. % downer cows in 12 months.
28. % mortality in 12 months.
29. Milk fever; % incidence/year.
30. Metabolic diseases; % incidence/year (such as ketosis or hypomagnesaemia, but not milk fever, mastitis or lameness).
31. Are sick animals kept isolated from healthy stock? (yes/no).
32. High incidence of coughing (yes/no).
33. Indicators of pain; methods and use of anaesthetics and analgesics for disbudding, dehorning and tail docking (description).
34. Subjective visual assessment of animals: use comparative drawings of healthy and content v sick and distressed cow.
35. Subjective auditory assessment of animals: grinding of teeth, bellowing, laboured breathing, coughing.

36. Subjective olfactory assessment of animals: odours of calves' breath, faeces, bedding, mouldy feeds.

6. Heat stress management

1. Average respiration rates (number/minute).
2. Frequency of cows with respiration rates exceeding 70 (%) and exceeding 100 breaths per minute (%).
3. Cow comfort index; proportion of resting cows standing or lying in free stalls (%).
4. Panting score (/5 points).
5. Cow comfort index; % cows standing or lying in free stalls.
6. Temperature Humidity Index (units).
7. Score on shed ventilation (/5 points).
8. Climate control equipment: fans, sprinklers, regular usage of hoses to cool cows.
9. Are stock fed during the cooler evening following very hot days? (yes/no).

7. Housing dairy stock

1. Leg score; (/3 points), % cows with severe rotation of their feet.
2. Cows with swollen hocks; % cows/year.
3. Cows with ulcerated hocks; % cows/year.
4. Cows with non-hock traumatic injuries; % cows.
5. Ease of movement in laneways; % cows slipping and falling.
6. Resting behaviour; % cows lying partly or completely outside resting area.
7. Social behaviour; number of head butts and displacements/cow/hour.
8. Flight distance, measured by approaching cows at the feed trough from a distance of 2.5 m and measuring the distance between the hand and muzzle at the moment the animal withdraws.
9. % of cows 0 to 10 cm, 10 to 50 cm, 50 to 100 cm and > 100 cm.
10. Cow hygiene scores for back of udder, side of udder, thighs and hips and legs and hooves (/4 points): cows with excessively dirty lower hind legs, hindquarters or udders (%).
11. Idle cows; % cows standing but performing no activity, such as rumination
12. Rumination; % cows standing or lying that are ruminating.
13. Rising restriction; % cows showing severe difficulty in rising, or hitting fittings as they rise or seen to be 'dog sitting'.
14. Cows with hair loss in lower limbs; % cows/year.
15. Score on manure and waste management (/5 points).

16. Subjective assessment of cleanliness of cow shed and on manure and waste management, such as daily removal of slurry and hosing down concrete floors.
17. Shed layout: cement v dirt floors, good v poor drainage, tie stalls v free stalls v open lot.
18. Adequacy of effluent disposal and storage system.
19. Natural ventilation: open shed sides, incidence of cobwebs in roof.
20. Is there sufficient shelter from extreme weather? (yes/no).
21. Resting area: open lot (resting area per cow), free stalls (type of bedding), tie stalls (length of tie, floor type), outside area for night resting.
22. Is there a minimum of dry lying area of 3.5 m^2 for adult cattle/bulls and 2.5 m^2 for growing heifers? (yes/no).
23. If permanently tethered, are cows and calves given access to move freely each day? (yes/no).
24. Adequacy and cleanliness of feed troughs and water containers/troughs
25. Adequacy of drinking water supplies; stock per drinker (number) or per cm of drinking trough (number/cm), water flow (L/minute), clean water not concentrate slurry.
26. Climate control equipment: fans, sprinklers, regular usage of hoses to cool cows.
27. Young stock housing: tie stalls, separate pens, cleanliness of floors.
28. Subjective olfactory assessment of shed: odours of air (hence ventilation), bedding and from poor drainage.
29. Resting behaviour: cows lying partly or completely outside resting area (%).
30. Social behaviour; number of head butts and displacements/cow/hour.
31. Do cows and calves approach stock people? (yes/no).
32. Free stall use: proportion of resting cows using free stalls (%).
33. Proportion of idle cows: cows standing but performing no activity, such as rumination (%).
34. Cows with stereotypic behaviour: such as pacing or swaying in tethered cattle, extreme grooming with the tongue, bar biting, tongue rolling, urine drinking or 'dog sitting' (%).

8. Producing quality milk

1. Are there routine checks made on subclinical mastitis, using the California Mastitis Test kit? (yes/no).
2. Average fat content (%).
3. Average SNF content (%).
4. Average total solids content (%).

5. Average total plate count or TPC (/ml); proportion with TPC exceeding 200 000, 500 000 and 1 million (%).
6. Methylene blue reductase test or MBRT (hours) or some other measure of bacterial contamination of raw milk.
7. Average somatic cell count or SCC (/ml): proportion with SCC exceeding 200 000, 500 000 and 1 million (%).
8. Score on milking sanitation (machine, facility, sanitation) (5 points).
9. Teat score (points out of 4): proportion of cows with very rough calloused teat ends (%).
10. Proportion of farm milk delivered to market (%).
11. Proportion of milk penalised for poor quality and/or composition (%).
12. Access to hot water pre and post milking.
13. Is detergent used during milking equipment sanitation?
14. Cleanliness of milk containers.
15. Are milk containers hung upside down to dry? (yes/no).
16. Cleanliness of teats and udders pre-milking: is each cow's udder washed with a fresh clean cloth? (yes/no).

7.6 Quantifying improved herd management

7.6.1 Assessing adequacy of the current herd management using cow milk yields

All the above KPIs can be quantified to provide guidelines as to which ones require priority in any dairy farm improvement program. Although some are relatively easy to quantify, others are quite difficult. Probably the most simple, and most used, single measure of SHD farm performance is the average daily milk yield of the milking cows. The correct term for this figure is 'rolling herd average' as it is the average milk yield of all the lactating cows, which will be at various stages in their lactation cycle.

This single value provides a summation of all the important aspects of SHD farm management, so any interpretation must take into account a diversity of feeding, herd and farm factors. Accordingly, many dairy specialists may query its usefulness as a single measure of dairy farm performance. However, it is routinely used by farmers to describe their farm's performance in relation to their neighbour's farm and also in relation to production targets provided by many government advisers. In addition, government officials often quote this measure when summarising the stage of development of their national dairy industries. Table 7.8 attempts to describe the adequacy of the farm's dairy farm management practices using the rolling herd average; other factors to consider are also listed.

Table 7.8. Interpreting the adequacy of dairy farm management from cow milk yields.

Range in average herd milk yields on tropical South-East Asian dairy farms

Herd milk yield (kg/cow/day)	Adequacy of dairy farm management practices
5	Very poor feeding and herd management and low genetic merit cows (or milking buffalo)
7	
9	Typical of many SE Asian smallholder farms, even with high grade Friesians
11	Gradual response with grade and crossbred Friesian cows to improved feeding, herd, young stock and shed management. *Milk yields of 15 kg/day are considered acceptable by many government dairy advisers*
13	
15	
17	
19	
20	Potential level in **lowland humid tropics** following improved management of body condition throughout lactation
25	High genetic merit cows in **tropical highlands** or **lowland dry tropics** with good farm management
30	Typical peak milk yields in herds with 25 kg/cow/day rolling herd averages
35	Unrealistic in SE Asia except where all major constraints to milk production have been overcome

Other factors to consider

- It is important to differentiate between rolling herd averages and peak milk yields.
- Milk composition is also a good indicator of feeding management.
- Excessive body condition is indicative of nutrient status and low production.
- Herd dynamics (number of dry cows and percentage of lactating adult cows in the herd) can also indicate adequacy of herd management.

7.6.2 An example of improved herd management in Indonesia

When comparing the herd response to improved management, controlled studies on research stations provide a better scientific environment to vary certain aspects of feeding and herd management while keeping others constant. However, farm-generated data is generally considered more realistic. To provide such a valuable dataset, a series of statistical analyses were undertaken on data generated over 10 years (1992 to 2002) to quantify such KPIs in both commercial farms and a government breeding station in Central Java.

Herd performance data from the breeding station (BS) were compared with data from nearby smallholder dairy farms (SHD) for Friesian cows of similar genetic merit with many variables, such as age of calving, parity and season of calving accounted for before comparing the location, or origin of the data (BS v SHD); 200 to 600 datasets per location were included in these analyses.

The authors (Anggraeni and Rowlinson 2005a,b) considered the BS herd to be intensively managed in contrast to the SHD's semi-intensive management. The increases in milk yield formed typically curvilinear patterns with the highest values recorded during the 3rd lactation. Parity and year of calving affected most of the lactation indices in Table 7.9 but not season of calving, indicating a sufficiency of forages to fulfil the needs of cows milking all year.

Table 7.9. Performance of cows managed in the breeding station (BS) or on smallholder dairy farms (SHD) in Central Java.

	BS	SHD
Lactation indicators		
Complete milk yields (kg)	4406	3287
Daily milk yields (kg/d)	14.5	9.8
Annual milk yields (kg)	3919	2731
305 day milk yields (kg)	4674	3096
Reproductive indicators		
Age at first calving (m)	29	31
Days dry (d)	104	96
Calving to 1st mating (d)	92	109
1st to 2nd mating (d)	52	45
Days open (d)	149	177
Calving interval (d)	418	433

Compared to the SHD cows, the BS cows produced more milk (averaging 4.7 kg/cow/day extra), were mated more quickly after calving (17 days), had fewer days open (28 days) and shorter calving intervals (15 days). They also first calved down 2 months before the SHD cows. However, they had slightly longer dry periods (8 days). Therefore, in herds of similar genetic merit, better herd and feeding management can increase milk yields by up to 5 kg/cow/day and reduce calving intervals from 14.4 to 13.9 months.

7.7 A checklist to assess current herd and farm management

The following series of observations can easily be made when visiting any dairy farm to assess the current management, performance and welfare of the stock and the likely profitability of the farm. Answers to the following questions could also be sought to help understand the current farm management.

7.7.1 Shed and facilities

- Roof height, eave height and natural ventilation.
- Temperature, humidity and air flow inside shed.

- Shed floor and cow lying area (cement, mats).
- Mats, enough for all cows, thin v thick.
- Stalls; tie stalls v free stalls v open lot.
- General stocking density and space for cows to rest.
- Adequacy and cleanliness of pens for young stock (heifers, milk-fed calves).
- Area for outside resting at night.
- Adequacy and cleanliness of feed troughs and water containers.
- Access to clean drinking water, not only as slurry feeding.
- Source and adequacy of water for drinking and cleaning.
- If sufficient cows, use of mechanical forage chopper.
- Room for feed processing and preparation.
- Adequacy and hygiene of milking area, including teat dipping and access to hot water.
- If machine milking, state of rubber linings.
- Cleanliness of milking buckets and milk cans.
- General hygiene and condition of cow teats and coats.
- Adequacy of effluent disposal and storage system.
- Adequacy of office and farm staff area.

7.7.2 Stock

- General condition (thin v good v fat).
- Obvious health issues, such as lameness.
- Is mastitis an issue? If so, what are the management procedures?
- Signs of heat stress (< 70 respirations/minute).
- % cows ruminating at rest.
- % of cows lying down and ruminating.
- Cow 'comfort' and contentment (obviously hungry and unsettled).
- Herd numbers and structure.
  - Milking cows.
  - Dry cows.
  - Heifers (weaning to calving).
  - Milk fed calves.
  - Other dairy stock (bulls steers).

7.7.3 Feed supplies

- Enough fresh forage fed each day? Typical amount fed per milking cow in wet/dry season.
- State of forage (improved v native v forage by-products, immature v mature).
- Sourcing fresh forage, grown v off farm.
- Enough concentrates fed each day? Typical amount fed per cow.

- Concentrates (formulated v mixed).
- What by-products are fed?
- If on farm mixing, specific feed additives (macro minerals, vitamins/minerals, rumen buffers).
- Use of shed effluent for forage production.

7.7.4 Answers to simple questions

- How many cows did you milk yesterday?
- How much milk did you sell yesterday?
- How much milk did you use to feed your milk-fed calves yesterday?

These answers should allow you to calculate the average milk yield per milking cow.

- How much per litre of milk were you paid yesterday?
- If there is a milk grading scheme, what grade was your milk yesterday?
- What was the typical composition of your milk (fat, solids not fat, protein).
- What was the typical quality of your milk (somatic cell count, TPC or MBRT).

7.7.5 Answers to more complex questions

- How aware are you of the importance of colostrum feeding to your newborn calves?
- Do you keep any farm records? If so, which ones?
- Is mastitis a problem?
- Is cow lameness a problem?
- What is the typical peak yield of your cows in early lactation?
- What are your typical number of days between calving to conception?
- What is your typical number of days between cows drying off and then calving.
- What is your typical age (number of months) of heifers when they first calve down?
- Are there animal health management and biosecurity plans in place? If so, describe them (protocols for biosecurity, mastitis, lameness, reproductive disease management, vaccinations, parasite control etc.).
- What is your typical proportion of calves that die during milk rearing?
- How many years have you been milking cows?
- Will you still be milking cows in 1 year's time or 5 years' time?
- Do your children want to follow you on the farm?
- Name three of your biggest problems on your farm. This can be any constraint at all, such as labour supplies, government/coop or milk processor support and services, dry season forage supplies. Poor milk returns and high cost of production are universal problems for all SHD farmers, so don't include them unless they are a real problem on your farm.

7.8 A checklist for assessing current calf and heifer rearing systems

More specific to young stock management, the following checklist can be used.

7.8.1 Peri-natal (Pre-calving and first 24 h)

- Pre-calving facilities and hygiene.
- Note the following observations soon after birth:
  - Calf is sitting in a sternal position within 5 min.
  - Calf makes standing attempts within 15 min.
  - Shivering often begins within 30 min.
  - Calf is standing by 1 hr.
  - Calf is suckling within 2 hr.
- Separation of calf from dam.
- Hours from birth when drinking first colostrum.
- Colostrum feeding (awareness of how best to obtain benefits from colostrum, putting theory into practice).
- Calf identification and recording.
- Housing of pre-weaned calves (stalls, bedding, separation from adult stock, cleanliness).

7.8.2 Pre-weaning feeding management

- Develop feeding programs to promote rumen development.
- Continual access to clean water.
- Type and method of liquid feed (type, feeding frequency, method of feeding, daily feeding rate).
- Type of concentrate (age when commence feeding, quantity fed, formulation).
- Type of forage (age when commence feeding, fresh v hay, forage quality).

7.8.3 Pre-weaning herd and health management

- Development of animal health program.
- Pre-weaning calf mortality.
- Condition of the milk-fed calves (body condition, signs of ill thrift).
- Prevention and treatment of scours (electrolytes, identification of causative agent, effective use of antibiotics).
- Prevention and treatment of pneumonia.
- Prevention and treatment of other animal health issues (such as lameness or climatic stress).
- Skills with identifying and nursing sick calves.

- Routine husbandry practices (disbudding, vaccinations, drenching, ectoparasite treatment, castration).
- Drug storage.
- Hospital pens.
- Veterinarian support (when sought, quality of support).
- Document pre-weaning morbidities (sickness) and mortalities (deaths).
- Fate of bull calves.
- Sources of labour (owner, family, employed and their efficient usage).
- Shed environment (effluent management, ventilation, coping with heat stress).
- General stock welfare and handling facilities/skills.
- Record keeping (system, feed inputs, animal health, labour).
- Calculating total pre-weaning costs.
- General hygiene (cleaning of feeding equipment, calf pen and shed hygiene).

7.8.4 Weaning process

- Basis of weaning (age, weight, concentrate intake).
- Movement to weaner pens.
- Reduce milk offered slowly or instantaneously.

7.8.5 Weaning to first calving

- Continual access to drinking water.
- Shed environment (effluent management, ventilation, coping with heat and cold stress).
- Feeding management (forages and concentrates).
- Chopping forages.
- Feed storage facilities.
- Monitoring feed intakes, weight, body condition, heat stress (stock and shed).
- Post-weaning heifer mortality.
- Condition of stock (body condition, signs of ill thrift).
- Age of first oestrus.
- Criteria for mating (age, weight).
- Mating program (bull v AI).
- Routine pregnancy testing.
- Preparation for calving (pre-calving facilities and hygiene).
- Calculating total rearing costs.
- Fate of weaned bulls.
- Age at first calving.
- Wastage rate from birth to 2nd calving.
- Subjective assessment of overall farm management skills.

7.9 Examples of improved management practices

The following lists some of the improved management practices that should be key objectives of farm development programs and should be demonstrated on model farms.

7.9.1 Sheds and facilities

- Loose housing rather than tie stalls.
- Free stalls rather than open lounging.
- Rubber mats in the free stalls.
- Outside sand yard for resting at night.
- Good slope on floor to aid cleaning.
- Flood wash system to clean floor.
- Adequate roof height.
- Cattle crush and yards for ease of handling stock.
- Mating yard if using bulls not AI.
- Young stock rearing area not directly adjacent to adult cow area.
- Individual pens or cages for milk rearing calves.
- Effluent dam to minimise nitrogen losses through volatilisation.
- Pump and pipes to distribute effluent onto forage production area.
- Mechanical milking, not hand milking, probably with bucket milkers.
- Hot water to clean milk handling and milk feeding equipment.
- Separate area in which cows can give birth.
- Hospital or isolation pen for sick stock.
- Vermin-free and insect-proof storage area for feeds.
- Refrigerator and lockable cabinet for drug storage.
- Fans and maybe even sprinklers for heat stress management.
- Plant trees and grass around sheds.
- Area for staff to relax when not working.
- Office area for keeping and storing farm records.

7.9.2 Feed and forage management

- Routine use of inorganic fertilisers on grass production area.
- Short harvest interval for Napier grass (30 days not 60 days).
- Consider routine wilting of freshly harvested forage to improve appetite, hence productivity.
- Consider making silage out of excess wet season forage to feed back to stock in dry season.
- If using raw ingredients for concentrate formulations, a separate area for mixing them on the floor.
- Ensure can source adequate supplies of quality forages to make up for any shortfalls in home-grown forages.

- Ensure can source adequate supplies of quality ingredients for supplementing forages.
- Provision of adequate fresh clean drinking water.
- Harvest and feed adequate quantities of quality green forages (say 40 kg/cow/d) to milking herd.
- Use mechanical chopper for processing forages.
- Consider ribbon mixer or small TMR wagon to mix all the feed ingredients.
- Consider routine analyses of nutritive value of all feed inputs.
- Formulate rations based on minimum costs of feed energy and protein within each feed type.
- Feed scales for weighing feeds.
- Chest girth tape for assessing stock live weights.
- Picture guides to routinely monitor body condition.

7.9.3 Herd management practices

- Feed colostrum immediately following birth.
- Consider calf milk replacer as cheaper alternative to whole milk.
- Ensure calves are fed high protein concentrate formulations, not the same as for milking cows.
- Develop regular vaccination for young stock (Clostridium) as well as adult stock.
- Monitor respiration rates to decide on heat stress alleviation procedures.
- Learn to identify early symptoms of ill health in calves, heifers and adult cows.
- Develop animal health protocols for milk-fed calves with local veterinarian.
- Develop animal health protocols for weaned heifers and adult cows with local veterinarian.
- Follow recommended Best Management Practices (BMP) for all aspects of herd management such as feeding, breeding, rearing young stock, milk harvesting, animal health and stock welfare.
- Do not use calves for milk letdown before milking.
- Only wash teats when preparing cows for milking.
- Ensure cows stand for 30 min following milking.
- Change milk liners after every 2500 milkings.
- Encourage staff to be involved in monitoring stock health and performance during their daily work routines, and writing observations down on a white board.
- Collect sufficient production and financial data to routinely monitor the relevant Key Performance Indicators.

Good cow performance depends largely on adequate intakes of high quality feeds as depicted in Figure 7.2 (molasses) and Figure 7.3 (leucaena) supplements.

Figure 7.2: In sugarcane-producing areas, molasses is a cheap, high energy feed for milking cows.

Figure 7.3: Leucaena leaves are a high quality protein, roughage source.

Figure 7.4: Permanent tethering of milking cows contravenes all our concepts of good cow welfare. Logic tells us that even a 1 or 2-hr break every day, when they are allowed out into a small yard next to the shed to 'stretch their legs' will stimulate their appetite, produce more milk (provided they are offered more feed) and make the farmer more money.

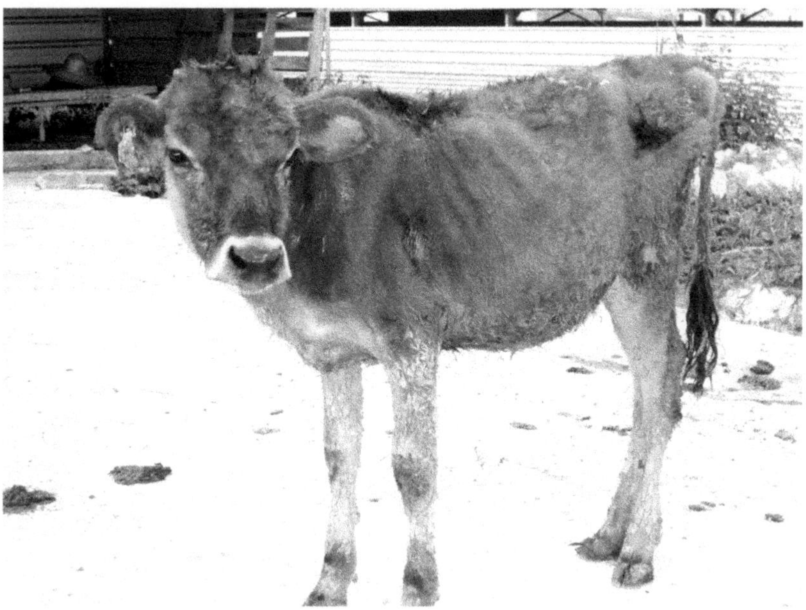

Figure 7.5: This 'half starved' Malaysian Jersey heifer is an extreme result of poor feeding management but it is all too common on many smallholder farms throughout the tropics. It will never be a productive member of the farm's milking team. Farmers must not look at heifer rearing as a farm cost; it is an investment in the farm's future.

However, if herd management is suboptimal, such as permanent tie stalls (Figure 7.4) and poor young stock rearing (Figure 7.5), cow appetites are depressed leading to reduced milk yields and fertility.

8

Best management practices on large-scale, intensively managed dairy farms

This chapter presents an overview of the essential on-farm practices necessary to effectively run a large-scale intensive dairy farm in the humid tropics to sustainably produce a consistently saleable product.

The previous chapter (7) discussed the Best Management Practices (BMP) specific to smallholder dairy (SHD) farmers. Many of these were grouped into a series of Golden Rules, for ease of developing farmer capacity building programs under the auspices of the Smallholder Dairy Development Program (SDDP). As the BMPs don't greatly differ for small or large-scale dairy farmers, this chapter will not repeat much of Chapter 7, only concentrating on those aspects that are particularly relevant to large-scale intensive dairy farming.

Section 7.1 in Chapter 7 on the importance of record keeping on dairy farms applies equally to large-scale farms as it does to smallholder farms, so should be read and practised in conjunction with the contents of this chapter. It is highly likely that management of large-scale dairy operations will have developed their own system of record keeping, particularly if they have the type of electronic cow monitoring system described in Section 5.6.2 in Chapter 5.

8.1 Forage production

As it is cheaper to grow quality forages on-farm, generally the less purchased, the lower the feed costs. With well-planned dairy production systems, it should be possible to supply 95 to 100% of the forages from on-farm supplies, through strategies such as fodder conservation. Strategic purchases of small quantities of very cheap, lower quality forages (such as rice straw) for stock with lower daily nutrient requirements, such as dry cows, may still be a good management decision.

The key issues to ensure high yields of quality forage from any forage crop are:

- Preparation of the soil for sowing, to ensure a good establishment of the crop and minimum weed invasion.
- Fertilising the crop with sufficient soil nutrients, using inorganic fertilisers to supplement dairy shed effluent. Ensure fertilisers are routinely applied after every (or at least every second) forage harvest.
- Developing a crop-harvesting program to ensure quality forage as well as adequate forage yields.

Fertilisers cost money, but they return more through improved yields and quality of forage, hence more milk. Provided other soil nutrients are not limiting plant growth, urea fertiliser can produce an extra 9 kg forage dry matter (DM)/kg urea or 18 kg DM/kg N applied. When harvested and fed to milking cows, this extra forage can yield an additional 9 kg milk/kg urea N (STOAS 1999).

One major limitation of forage production on most dairies throughout tropical Asia is the poor adoption of inorganic fertilisers. Use of cow manure only to fertilise grasses is common practice in most dairying areas in Asia because many farmers are not even aware of the economic gains achievable through using inorganic fertilisers. Cow manure supplies organic matter to the forage area, but insufficient N to maximise forage yields and quality. Dairy farmers should apply at least 100 kg urea/ha/yr to their forage production area, in addition to the recycled manure.

The general recommendations for moderate milk yields (10 to 15 kg/cow/day) in 450 kg dairy cows consuming 13.5 kg DM/d, is for Napier grass containing 8 to 13% protein and 7 to 8 MJ/kg DM of metabolisable energy (ME). This can be supplied from forage harvested at 42 to 70 days, when 60 to 100 cm high, providing a carrying capacity is 4 to 4.5 cows/ha. For high milk production (< 15 kg/cow/d), harvest intervals would have to be increased to 30 days. Milk yields per ha would be high even with low carrying capacity, because of the reduced protein yields. The increased cost of more frequent harvestings and greater fertiliser applications to maintain soil fertility will reduce the economic benefits of the higher milk yields.

8.1.1 Stocking capacity

For a farmer growing the maximum quantities of quality forages, to feed his milking cows well, he should have no more than eight to 10 milking cows per hectare of forage grown on his farm. The derivation of these stocking capacities can be found in Chapter 20 of Moran (2005). However, most dairy farmers do not manage their forages well enough to produce the highest yields of forage. Therefore, a more realistic recommendation would be six to eight milking cows (plus the replacement heifers) per hectare of forage grown on-farm. Figures 8.1 to 8.3 show forage maize arriving at a dairy, the forage being put through a chopper and farmers examining the feed after processing.

8.2 Feeding management

8.2.1 Forage quality

To produce milk and calves, dairy cows require feed nutrients that are supplied through a combination of forages and concentrates. To produce acceptable milk yields, say 15 L/day, cows require a ration containing at least 10 MJ/kg DM of

Figure 8.1: Forage maize, contract grown by local farmers, being delivered to the North Vietnam dairy feedlot for ensiling.

Figure 8.2: Processing the North Vietnamese forage maize through a chopper before consolidating it in a large silage bunker.

Figure 8.3: Checking the consolidation of the maize greenchop in another Asian dairy feedlot.

metabolisable energy (ME). The more ME that is supplied by forages, the less is required by concentrates. For milking cows, the recommended forage quality would be 9.5 to 10.0 MJ/kg DM of ME and 12 to 14% crude protein.

8.2.2 Concentrate feeding program

Concentrates should be formulated to provide adequate dietary nutrients to supplement available forages. The recommended concentrate quality would be 11 to 12 MJ/kg DM of ME and 16 to18% crude protein (CP), depending on milk yields.

Many Asian dairy advisers use a general 'rule of thumb' that farmers should feed 1 kg concentrate for every 2 L of milk produced. This is a safety measure because of the lack of knowledge on the nutritive value of the feeds, particularly the forages. It also provides supplemental energy to cows when fed only limited amounts of forage.

With knowledge of the feeding value and costs of the forages and concentrates, more objective and therefore better decisions can be made on how much concentrate should be fed to achieve target milk yields. This requires more knowledge and greater effort than following the 'feed 1 kg concentrate per 2 L milk' rule, but such decisions can greatly reduce total feed costs and so improve profitability, when expressed as milk income less feed costs.

8.2.3 Total feed costs

The choice of available feeds for milk production will differ from country to country, as do their relative costs. The principle of formulating profitable rations is to compare different feeds first on the basis of their cost per unit energy, because energy is nearly always the first limiting nutrient. When protein deficiencies limit cow performance, the unit cost of protein becomes important. The fibre content of each potential feed ingredient is considered just to make sure the cows' appetite will not too restricted and they will not eat all that is offered.

When formulating rations, either by computer or by calculator, as long as farmers are confident that the raw data (feed costs and nutritive values) are representative of that feed being fed to those cows, then traditional ration formulation calculations, such as those presented by Moran (2005), will provide a meaningful answer to any least-cost ration. With experience, the process is less time consuming because 'best bet' rations can be easily checked for their nutrient content and likely milk yield response.

8.2.4 Milk income less feed costs

Milk income less feed costs (MIFC) is one of the simplest indicators of farm profitability. In addition, changes in MIFC are quick to monitor because of the rapidity with which milking cows respond even to small variations in their feeding management. When introducing new feeds into the diet or varying their amount, the cows' milk responses will reflect these changes within a few days as will their

MIFC within a week or two. The development of generic indicators for total feed costs and MIFC depend greatly on the base costs of feeds in different dairy regions. MIFC is also known as Feeding Profit and to aid its calculation, the senior author has developed a computer program called INDOFEEDPROFIT, which will be discussed in Chapter 11, Section 11.4.2 of this manual.

8.2.5 Pattern of milk production

The three major factors determining total lactation milk production are the lactation length, the peak lactation yield (within 6 to 8 weeks post calving) and its rate of decline from this peak (or lactation persistency). Persistency will be discussed in Section 8.7.5 of this chapter.

8.3 Young stock management

Poor heifer management is a major problem in many (if not most) Asian dairy farms (both large and small). Farmers give insufficient attention to young stock because they do not generate income for many months until after they first calve down. In addition, the first three months are the most expensive period in the life of any dairy cow and many farmers are just not prepared to invest in the calves' future. A low calf mortality rate indicates that early milk rearing practices are adequate, hence will provide greater opportunities for economic and genetic improvement in the herd.

There are easily quantifiable benefits in having more newly calved heifers available to replace older unprofitable cows, as heifer and reproductive managements improve. These benefits are for:

- every month reduction in age at first calving; 1 to 2% more first calf heifers
- every 10% reduction in calf mortality; 3 to 5% more first calf heifers
- every month reduction in inter-calving interval; 2 to 3% more first calf heifers.

Farmers should aim to rear 20 to 25% of their milking herd each year as replacements, to calve down for the first time by about two years of age and produce at least five calves during their productive life. Realistic targets for Friesians in tropical dairy systems are:

- Calf mortality to weaning, 4 to 6%.
- Heifer wastage rate from birth to second calving, 20 to 25%.
- Live weight at mating, 250 to 300 kg.
- Live weight at first calving, 400 to 500 kg (depending on breed type).
- Age at first calving, 24 to 30 months.

Another good indication of heifer management is first lactation milk yield, expressed as % of mature cow production, with a target of 80 to 85%. If this is less than 75% of the mature equivalent, then the heifer-rearing program should be reviewed.

Wither height (or height at the shoulder) is a good measure of bone growth and potential body frame size in heifers. Frame size can influence ease of calving and appetite of milking cows. For Friesians, farmers should aim for wither heights of 115 to 120 cm by 15 months and 125 to 130 cm by 24 months of age.

8.3.1 Caring for the newborn calf

There are well 'tried and tested' procedures for managing newborn calves. These include:

- Making sure the calf is breathing by clearing the nose and mouth. If breathing is slow to occur, stick a piece of straw up the nose to stimulate breathing and if this fails then pour cold water over the calf's head, especially into its ear canal.
- Removing the calf from any contaminated bedding by providing clean bedding where the dam can lick the calf.
- Placing the calf into a dog sitting position to encourage breathing and disinfect the navel cord with 7% iodine solution at least once.
- The next steps can be summarised in three statements: *Get her up, get her dry and get her fed.*
    - *Get her up.* Encourage the calf to stand.
    - *Get her dry.* Allow the dam to lick the calf and/or vigorously rub the calf with a dry towel to create a dry fluffy coat.
    - *Get her fed.* Assist the calf to suckle from a clean teat so it can immediately drink colostrum.
- Minimise the possibility of pathogens entering the calf via two major routes, namely the mouth and the navel.
- Permanently identify each calf.
- Calves born from difficult calvings should be clearly marked to closely watch over the next few weeks.

In summary, for a normal delivery, a newborn calf should exhibit the following signs:

- The calf tries to lift its head within minutes.
- It rolls up on its sternum within 5 min.
- It attempts to stand within 15 min.
- It is standing within 1 hr.
- It suckles within 2 hr.

If calves do not exhibit these signs, chances are that the delivery is not normal and the calf requires assistance.

8.3.2 Managing the milk-fed calf

There is no single best way to milk-rear calves. All sorts of combinations of feeding, housing and husbandry can be successful in the right hands and on the right farm.

The essence of good calf rearing depends on two major nutritional factors. First, an adequate intake of high quality colostrum within the first day of life and, second, feeding management to encourage early rumen development.

Colostrum feeding

Calves are born with no immunity against disease. Until they can develop their own natural ability to resist disease, through exposure to the disease organisms in their surroundings, they depend entirely on the passive immunity acquired by drinking colostrum from their dam.

Colostrum is the thick, creamy yellow, sticky milk first produced by cows initially following calving, and contains the antibodies necessary to transfer immunity to their calves. It is essentially milk reinforced with blood proteins and vitamins. It has more than twice the level of total solids than whole milk through boosted levels of protein and electrolytes. It also contains a chemical allowing newborn calves to utilise their own fat reserves to immediately provide additional energy.

The chances of calves surviving the first few weeks of life are greatly reduced if they do not ingest and absorb these antibodies into their bloodstream. It takes far fewer disease organisms to cause disease outbreaks in such calves than if they can acquire immunity from their dam. Calves without adequate passive immunity are four times more likely to die and twice as likely to suffer disease, than those with it. Furthermore, in certain situations, blood levels of antibodies in heifer calves are directly related to their milk production in later life.

There are three key principles for colostrum feeding which can be summarised as the three Qs, namely:

- *Quality.* Only provide good quality colostrum.
- *Quantity.* Ensure calves ingest sufficient antibodies.
- *Quickly.* The timing of the first feed must ensure efficient absorption of the antibodies into the blood.

In summary, the important principles of good colostrum management are:

1. Use colostrum from mature cows that produce less than 8 L at their first milking.
2. Use only first milking colostrum, do not sterilise it but ensure the feeding equipment is clean and has been sterilised.
3. Feed 4 L to large calves or 3 L to smaller calves at first feeding.
4. Feed colostrum as soon as possible, at least within the first 3 hr after birth.
5. Do not let calves suckle their dams.

In summary, calves should have access to 4 L of colostrum within the first 6 hr of life. They may not need any additional milk for the next 12 to 24 hr. Any calf that is suspected of not having suckled in the first 3 to 6 hr should be hand-fed the colostrum. With sick or weak calves, colostrum may have to be administered by stomach tube. It is not difficult to stomach-tube young calves.

The longer calves spend with their dams, the greater their exposure to potential disease. Therefore, newborn calves should be separated from their dams as quickly as is practically possible.

Feeding milk to the calves

Calves require a diet of good quality whole milk or calf milk replacer (CMR). The choice of liquid feed should be based on its cost per litre and this depends on the milk returns for raw milk and the cost of the CMR powder. In many cases, CMR is the cheaper when diluted to a mixture of one part CMR and 10 parts water (or as per the manufacturer's recommendations). The CMR must be a good quality product to ensure calves can digest it properly.

Milk can be fed either through teats or directly from a bucket. There is no difference in the ability of the calf to utilise the milk fed either way (Moran 2012b). If calves are sucking each other, teat feeding may be necessary to control this. Milk can be fed once or twice a day as it clots in the calf's stomach after which it is slowly broken down by the calf's gastric juices. The milk does not have to be heated up before feeding. It is important that the milk concentration, timing and temperature remain consistent from day to day. Calves should be given free access to good quality drinking water from 1 week of age.

It is very important that calves are kept in clean pens and that hot water is available to clean milk-feeding equipment. One major health problem is scours, which results mainly from an inconsistent milk feeding program, poor hygiene and inadequate colostrum feeding. As many scours are not due to bacterial diseases, there is often little need to consider antibiotics in the treatment. Fluids and electrolyte replacement are the main requirements.

8.3.3 Other aspects of calf management

Try sleeping on concrete

Concrete is always hard with no soft spots in which calves can get comfortable. Calves lie down a lot, spending 95% of their time when newborn and slowly decreasing as they approach weaning age. We must manage stress associated with lying comfort by using enough bedding to buffer the typical hard pen base. Even in the tropics, newborn sick calves can suffer from cold stress and concrete transmits body heat very easily. Clearly calves need a clean and comfortable place to rest. Wooden slats are easy to clean but ideally they need some form of insulated bedding, such as straw or sawdust. In cold conditions, calves need to be kept dry, well ventilated with fresh air, no wind and be able to nest in long straw. Many calf-rearing systems have high ammonia levels close to the ground, so get your nose to ground level to check this, even if you think the ventilation is good.

The calf kitchen

A dedicated place, the calf kitchen, to clean and store all the milk feeding equipment will improve the efficiency and hygiene of managing the young calves. This includes storage of calf concentrates and roughage and, if being used, storage and formulation of calf milk replacer. All milk-handling containers should be rinsed with cool water, washed with warm soapy water, rinsed again then hung upside down to drain. The kitchen should include hot water, clean rinsing water, a sink and drying racks. If using fresh milk, the kitchen should be within easy access of the milking area.

Hot weather and calves

Calves subjected to hot weather change their behaviour by decreasing feed intake, increasing water intake, increasing time spent standing and of course respiration rates. Calf health is also affected by heat stress, such as rapid dehydration, while vaccines work less well or not at all and their immune system defences against infection are weakened. The first response is to drink more water, which increases by 30% as temperatures increase from 21 to 27°C and by 100% as they increase from 21 to 32°C. Calves with scours may suffer extreme dehydration very quickly. So we need to be extra vigilant in diagnosing and treating these calves.

To reduce the effects of heat stress, it is important to move as much air around the calves as you can using natural ventilation and fans. Leave the fans on all night to help remove excess body heat. Drinking containers should be kept clean and free from algae and mould.

Only handle the calves during their low heat-stress times, such as very early in the morning. The calves' peak body temperatures occur between 4 to 7 p.m., when we perceive the day to be cooling off.

Calves can tell you how they are feeing

If you are 'listening' with your eyes, sick calves can clearly communicate that they are not feeling well. The most common sickness behaviours are:

- Smaller flight zones. Sick calves will not instantly stand up or move away when approached and may even need more encouragement to get up.
- Less interest in people. Calves that receive their milk from people should display an eagerness to interact with them when they are approached. Calves that fail to take at least one step forward in the first 30 s after being approached are more likely to be ill.

- Posture changes. Standing with an arched back and tucked in tail for at least 60 s after it has risen, stretched and urinated is a demonstration of physical discomfort and possible ill health.
- Calf behaviours should be observed when they are relaxed because if they are startled or frightened, their survival instincts will override their feelings of sickness and make ill calves more difficult to detect. Watching for these signs every day will help to identify disease incidence early so treatment and supporting therapy can be delivered promptly and effectively.

Recognising sick calves

Other tips for sick calf identification include:

- Watch for calves that are drinking slower or not finishing their milk or milk replacer solution.
- Watch for calves spending more time lying or resting or eating less calf concentrate.
- Rapid breathing, fever, watery eyes, diarrhoea, signs of dehydration, nasal discharge or coughing are signs that a health challenge has already developed.
- The degree of dehydration can be assessed using the skin fold (pinch) test. Pinch the skin and note how long it takes to return to normal. In healthy calves, this is less than half a second whereas it can vary from 1 to 10 s in sick calves, depending on their degree of dehydration. Another indicator is the degree of sunkeness of the eyes.
- Look for calves doing something that all the other calves in the pen are not doing. Healthy calves should be bright and alert, active and energetic, eating, chewing and playing.

It takes more than just milk

When thinking of calves and what it takes to grow them properly, the main component of their diet is milk or milk replacer solution. However, it is important to think about the other diet components as well. Milk is fed to calves for only a limited period and the transition from milk to solid feed is relatively quick. To create a smoother weaning period and maintain growth post-weaning, it is important to encourage intake of more than just milk, even in the pre-weaning period.

The main question is what else should we be feeding the milk-fed calves? Concentrates only or forages as well? Given the choice, calves tend to prefer forages before weaning (as long as they are good quality and palatable) and concentrates after weaning. However, after weaning, a combination of concentrates and forages promotes the best growth rates.

Early rumen development

The rumen is non-functional in newborn calves; hence, all digestion must take place in the abomasum (or true stomach) and the small intestine.

The weaned calf needs a fully functional rumen in order to be well adapted to a forage-based diet.

Rumen development occurs through the digestion or fermentation of feeds (roughages and concentrates) by the rumen microbes. Calves should be encouraged to eat solid feeds at an early age, mainly through limiting their access to milk to 4 L/day. From the first week, roughage such as clean straw should be offered in combination with high-quality concentrates specially formulated for rearing calves. But concentrates are more important in developing rumen function.

A program for milk feeding

If calves are strong, healthy and kept warm and dry, they can be successfully reared on once daily feeding with 4 L of whole milk, or its equivalent in milk replacer. All calves should be offered a specially formulated calf meal from one week of age. Milking cow concentrate formulations do not contain sufficient protein to meet the needs of young calves.

All calves must be given the opportunity to nibble on the straw even though they will eat very little of it. If good quality hay is fed, it should be limited to 100 to 200 g/calf/day. Calves should have limited access to fresh forages.

Feeding milk only once each day helps the calves to develop an appetite for the concentrates. It is the concentrate rather than the milk that should provide the bulk of nutrients to keep the calf growing. Calves can be weaned off milk once they are consuming 0.75 kg/day of concentrates for two or three consecutive days. This usually occurs by about 6 to 8 weeks of age.

Ideally, young calves should be housed individually or in small groups. They should also be individually bucket-fed. There is no advantage in milk feeding using teats rather than buckets; it only creates extra work in keeping them clean.

Weaning age

The age when calves should no longer be fed milk also depends on the quality of feeds available. They should be weaned at:

- 2 months, when quantity and quality of roughage and concentrates are good
- 4 months, when quantity and quality of roughage and concentrates are average
- 6 months, when quantity and quality of roughage and concentrates are poor
- 8 months, when suckling and cows are dried off.

It must be stressed that the cost of milk feeding, either the loss of income from these milk sales, or the cost or purchasing and mixing calf milk replacer, must be considered when considering delaying weaning age.

Concentrate quality

Milk-fed and weaned calves require concentrates containing higher protein levels (18–20%) than do milking cows (16%). Low protein concentrates will not promote the same rate of rumen and body development in milk-fed calves. Consequently weaning such calves at 2 months old may be too early.

The high cost of milk rearing

Many Asian farmers feed milk (or milk replacer) to their calves for 8 to 10 to 12 weeks. This is the 'accepted method' because the high nutrient intakes in milk should ensure a good growth rate, provided the calves remain healthy. This method is common where specially formulated calf concentrates are not readily available.

However, feeding high levels of milk can cause several problems. First, it is invariably more expensive to provide the same intake of feed nutrients from milk or milk replacer than from formulated calf concentrates.

The second problem is the increased likelihood of disease. Once milk is removed from their diet through weaning, calves are more resistant to scours. Unless a strict cleaning and sterilising routine is enforced in the calf shed, flies and other disease-carrying agents will thrive on residual milk left in buckets, on floors and in other equipment used with milk feeding.

8.3.4 Management of weaned replacement heifers

Weaned growing heifers require less attention than milk-fed calves and milking cows. From weaning until breeding and sometimes even after then, daily contact is not necessary. Because their nutrient requirements are relatively low compared to milking cows, heifers may be located away from the dairy farm, sometimes on agistment on other farms. Unfortunately, the saying 'out of sight, out of mind' applies too frequently to replacement heifers. This relative neglect is understandable in view of the long time it takes before any inadequacies in post-weaning practices are reflected in poor milking cow performance.

Dairy heifers need to be well fed between weaning and first calving. Growth rates should be maintained, otherwise heifers will not reach their target live weights for mating and first calving. Undersized heifers:

- have more calving difficulties
- produce less milk
- have greater difficulty getting back into calf during their first lactation
- when lactating, they compete poorly with older cows for feed
- because they are still growing, they will use some of their feed for growth rather than for producing milk
- they are more likely to be culled for poor milk yield and/or infertility.

The onset of puberty is related to weight rather than age. A delay in puberty means a later conception, which can disrupt future calving patterns and increase rearing costs. All heifers should reach a minimum weight before joining, as lighter heifers have lower conception rates. Target growth rates should be 0.6 to 0.7 kg/head/day.

Although replacement heifers are essentially non-productive animals, some expenditure is necessary. They represent capital and investment in the dairy herd's future. Heifer rearing should achieve the maximum return on this investment with a minimum of outlay. It should not be regarded as a haphazard undertaking, which hopefully will produce a pregnant heifer, but rather as a business enterprise with clearly defined goals such as:

- the number of animals to be reared
- their desired age at first calving
- their target live weight at calving
- their feeding program
- ways to monitor their performance and total rearing costs
- any specific housing and health requirements.

When rearing dairy replacement heifers, producers should have five major objectives:

1. The maintenance or expansion of herd size. Heifer rearing systems should provide sufficient animals to replace cows culled from the milking herd and allow for increases in herd numbers if required.
2. Calving by 24 to 30 months of age. Entry into first lactation by 24 months of age minimises the total non-productive days and maximises lifetime productivity.
3. Sufficient growth for minimal dystocia (that is calving difficulties) at first calving. Heifers need to be large enough to calve without difficulty.
4. Maintenance of health. The prevention of clinical and subclinical disease plays a large role in the ability of replacement heifers to meet live weight and age targets at first calving. Longevity and lifetime productivity is also affected.
5. Genetic progress. Replacement heifers generally have higher genetic merit than the current milking herd. This can be expressed as increased productivity (both milk volume and solids), improved efficiency of production and/or enhanced resistance to disease.

Heifer rearing is not cheap and the costs to produce a lactating first-calf heifer can account for 15 to 20% of the total milk production costs. It is not good economics to cut back on heifer-rearing costs as lifetime profits will be reduced. The costs of heifer rearing are discussed in Chapter 11 at Section 11.4.3.

The number of first calving heifers each year will depend on the replacement rate within the milking herd. This is the sum of the wastage rate caused by infertility, lameness, mastitis, low milk yield, old age, accidents etc., together with the particular voluntary culling policy for that herd, whether this is to improve milk yield, feed efficiency or reduce calving interval. The number of heifer replacements to be reared also depends on mortality rates during rearing, the conception rates at first mating and the proportion of heifers reaching target weight for ages.

With two to three heifers per 10 cows introduced into the milking herd annually, at least 80% of the milking cows should be mated (either using a bull or artificially insemination) to obtain that number of replacements each year. When determining the total number of calves to rear, consideration could be given to rearing additional heifers for sale to other dairy farmers and/or bull calves for dairy beef.

On well-managed farms in temperate dairy regions, achieving a consistent calving program, requires heifers to:

- reach puberty at ~12 months of age
- become pregnant at 14 or 15 months of age
- calve at 24 months of age
- return to oestrus and be mated within 70 to 80 days of calving.

8.3.5 In summary

Benefits of heavier heifers

Provided heifers are at least 18 months old, the younger the heifers calve, the higher their first lactation and mature milk yields, the more calves they produce and the longer their productive life in the herd. An additional benefit is a more rapid generation interval, and hence a faster rate of genetic progress in the milking herd. Lifetime productivity reaches a peak in heifers calving at 25 to 27 months of age. For example, Friesian heifers calving at 24 to 27 months of age can produce 21 000 L milk over a 7-year productive life on established temperate dairy farms, compared to 18 750 L if calving at 30 to 33 months and only 17 000 L if calving at 36 to 42 months of age.

Several long-term studies have documented the long-term benefits from heavier calving weights in Friesian heifers. For every additional kg at first calving, heavier heifers can produce 7 L extra milk in each of their first three lactations. Therefore, if heifers calved at 500 kg compared to 450 kg, they would produce an extra 350 L milk/lactation or 1050 L extra milk over their first three lactations.

Calves that are poorly managed before and after weaning are disadvantaged for their entire life. Even if they are well fed after mating, their ultimate mature size is restricted and if they do put on extra weight, it tends to be fat. Most of the growth in skeletal size occurs before, not after puberty.

Dairy heifers need to be well fed between weaning and first calving. If growth rates are not maintained, heifers will not reach their target live weights for mating and first calving.

Undersized heifers have more calving difficulties, produce less milk and have greater difficulty getting back into calf during their first lactation. When lactating, they compete poorly with older, bigger cows for feed. Because they are still growing, they use feed for growth rather than for producing milk.

Poorly grown heifers do not last long in the milking herd. They are more likely to be culled for poor milk yield or poor fertility during their first lactation.

Targets for replacement heifers

Traditional target weights are too low to ensure first lactation heifers achieve their productivity potential, particularly on farms where milking cows are well fed. Friesian crossbred heifers should weigh 280 to 340 kg with wither (shoulder) heights of 115 to 120 cm at mating then 450 to 500 kg with wither heights of 133 to 135 cm at calving.

Feeding heifers to achieve target live weights

Producers should regularly weigh or monitor their young stock, and then vary feeding strategies according to their growth rates. Growth in Friesians should average 0.6 to 0.7 kg/day, although that can vary between 0.5 and 1.0 kg/day, depending on available forage and the supply and cost of suitable supplements.

As fresh forage is the cheapest feed, it should constitute the bulk of the diet, with silage or hay and concentrates used to overcome forage shortages. Until calves reach 200 kg in weight, they are not able to maintain the growth rates needed to reach target weights on diets of most forages. Their capacity is limited and they simply cannot eat enough forage to meet their nutrient requirements for rapid growth.

Forage quality and allocation should allow for continuous growth throughout the first two years. Uniform growth is not necessary and may be impracticable with seasonality of quality forage supplies. However, heifers should not be allowed to lose weight or to grow very slowly for long periods of time.

8.4 Disease management

The three biggest animal health problems on Asian dairy farms are lameness, mastitis and scours in milk-fed calves. Treatment, control and management practices for these disease problems have been discussed in detail by Moran (2012a).

8.4.1 Health and reproductive performance

There are many health management problems that can adversely affect fertility, such as:

- management of twin calves
- assisted calving
- Retained Foetal Membranes (RFMs)
- uterine infections and vaginal discharges

- lameness
- ketosis
- displaced abomasum
- cystic ovaries
- abortions.

Some health problems affect the reproductive tract directly (such as RFMs and vaginal discharges) while others reduce feed intake leading to rapid body condition loss and anoestrus (such as lameness and ketosis). It is important to keep good records and have a planned approach for treatment and prevention.

Cows with problems at calving have an increased risk of infection of the reproductive tract. Such infections can last for weeks (even months) after calving and can even show normal heats and no abnormal discharge, yet these cows can have reduced fertility. They may cure themselves over time but are more likely to show repeat heat cycles. As well as immediate treatment, follow-up treatments are available such as prostaglandin or antibiotics.

To improve cow health, it is important to keep accurate records and seek veterinary advice if the percentage of naturally calving cows (excluding cows induced to calve early) with this problem exceeds the number shown in brackets below:

- any cow having twins (no practical strategies prevent twin calves)
- any assistance required to deliver a calf (seek advice if > 6%)
- any calf born dead or died within 24 hr of birth (seek advice if > 1%)
- RFMs, that is membranes visible externally on the day of calving (seek advice if > 4%)
- vaginal discharge or pus discharging from the vulva more than 14 days after calving (seek advice if > 6%)
- lameness, or cows not bearing full weight on at least one leg which affects walking (seek advice if > 3% of first calvers or > 2% of older cows)
- abortions (seek advice if > 5%)
- other health problems, including ketosis, displaced abomasum or cystic ovaries (seek advice if > 5%).

Some infections that cause abortions in cattle can also infect humans, so:

- only handle aborted foetuses and membranes with disposable gloves
- avoid contact with vaginal discharges from aborted cows
- bury the foetus and membranes ensuring that dogs are not allowed access to them.

8.4.2 Biosecurity when purchasing new stock

Most dairy farmers purchase new stock during the farm establishment process. Therefore, it is important to carefully plan the introduction of stock to minimise the risk that they will introduce infectious diseases. The factors shown below are important in reducing this risk.

When purchasing new stock:

- ensure their health status is known
- where possible, ensure details of their vaccination program are known
- avoid purchasing stock from unknown sources or stock that have mixed with other cattle before sale
- purchase heifers because they can be more easily quarantined and are less likely to have mastitis
- calves purchased should be kept separate for at least a week with no signs of disease
- transport purchased cattle preferably in the farmer's vehicle, in a clean truck or trailer
- quarantine them in an area separate from other cattle on the farm
- use a medicated footbath before allowing them to enter the herd
- vaccinate them during the quarantine period to make sure they are integrated into the farm vaccination program.

Other biosecurity measures that should be considered include:

- controlling the movement of people, animals and equipment onto the farm
- as some diseases are spread on clothing and boots, if equipment is borrowed from other farms, it should be cleaned before use
- keep transport or service personnel away from the main herd area, especially the calf shed, or at least provide them with footwear.

8.5 Reproductive management

8.5.1 A framework for improved fertility

There are six factors that have large influences on herd reproductive performance. Three are non-nutritional and three are nutritional. They are:

1. The length of the *voluntary waiting period*, that is the number of days delay after calving before farmers begin inseminations. This is 50 to 55 days in herds with the best fertility.
2. *Heat detection.* Farmers can make two types of mistakes: they can diagnose heat in cows not on heat (called a false positive) or miss a heat identification (undetected heat). Missed heats are more common. The higher the heat detection rate, the higher the submission rate. Farmers with over 80% heat detection rates had 73% 80-day submission rates.

3. *Artificial insemination* (AI) practices. Good first insemination rates were 45 to 48%.
4. *Body condition.* Cows calving at condition scores of 3 (where 1 = emaciated and 5 = extremely fat) had higher 100-day in-calf rates (54%) than those calving at less than 2 (41%). Cows calving in very high score (4 or more) may lose condition more rapidly after calving and can suffer reduced fertility.
5. *Feed intake.* Better-fed cows have higher fertility. Better feeding can improve 100-day in-calf rate from 41 to 57% and reduce 200-day not in-calf rate from 15% to 9%.
6. *Heifer live weight.* The occurrence of the first oestrus in yearlings depends on live weight. Therefore, better feeding practices in early life will lead to younger age at first calving in virgin heifers. These heavier animals will also cycle earlier after calving.

For year-round calving herds, there are four useful measures of reproductive performance. These are detailed below.

100 day-in-calf rate

This calculates the percentage of the cows in the herd that become pregnant by 100 days after calving. It also describes how many cows will calve within ~13 months of their previous calving. High 100-day in-calf rates lead to fewer cows with long intervals between calving and fewer cows culled as non-pregnant. Cows that conceive within 100 days of calving will calve again within 12.5 months and generate higher profits than cows that take a longer time to conceive or fail to get pregnant. This measure usually allows for the voluntary waiting period (the days between calving and the first mating) of say 55 days plus two oestrous cycles of AI (say 42 days) before the cow is put out for natural mating.

200-day not-in-calf rate

This calculates the percentage of cows not pregnant by 200 days after calving. Farmers want as many cows as possible to calve no more than 15 to 16 months after their previous calving. This coincides with six months after which non-pregnant cows are often culled. It cannot be calculated until many months after cows have calved, but because it is closely related to the 100-day in-calf rate, it can be estimated from that measure. It cannot be calculated unless the whole herd is pregnancy tested.

Submission rate

This is the percentage of the herd that received at least one insemination within a specified number of days after calving. To achieve a high 100-day in-calf rate, a high percentage of cows in the herd must be submitted to insemination with minimum delay after calving. An 80-days submission rate is the percentage of cows that receive at least one insemination by 80 days after calving.

Conception rate

This is the number of services resulting in pregnancy divided by the total number of services. This describes the percentage of inseminations that are successful and result in pregnancy. This has always been considered an important measure of reproduction but it does not fully describe overall herd performance. Herds can have high conception rates but poor 100-day in-calf and high 200-day not-in-calf rates. Sometimes the first insemination conception rate is calculated by including only the first services after calving in the analyses.

For a milking herd in Asia, target KPIs for reproductive performance are:

- 100-day in-calf rate; 55 to 60%
- 200-day not-in-calf rate; 13 to 15%
- submission rate; 65 to 70%
- voluntary waiting period; 50 to 60 d
- conception rate to first insemination; 45 to 50%
- inseminations per conception in an AI program; 1.8 to 2.0.

These measures of reproductive performance are rarely used in Asia, because they require routine pregnancy testing of the entire herd. More typical ones are:

- days from calving to first service; 60 to 80 d
- inter-calving interval; 12 to 13 months.

8.5.2 Combined measures of reproduction and calf survival

Table 8.1 presents measures of reproduction and calf survival in two rearing systems, to calculate their relative replacement rates for a dairy herd with stable stock numbers. System A measures could be considered as a set of key indicators.

Table 8.1. Measures of reproduction and calf rearing to produce replacements for a stable dairy herd.

Rearing system	A	B
Calving interval (m)	12	18
Calving rate (%)	85	65
Stillborn calves (%)	2	5
Calf mortality from 0–24 m (%)	8	20
Non-pregnant heifers (%)	5	10
Heifer calves born (%)	36	15

Assuming cows remain in the milking herd for four to five lactations, 20 to 25% should be replaced each year. From Table 8.1, the supply of 36% heifers from System A allows for the sale of young breeding stock or a higher culling rate to better address genetic improvements in the herd. Only one in every six or seven cows could be replaced annually in System B, which would hardly be enough to maintain herd numbers, let alone allow for much genetic selection.

With high ages at first calving (> 30 months) and long inter-calving intervals (> 15 months), it is very difficult to increase herd size through natural increase. That is why it is so important to seek the underlying causes of herds with high percentages of dry cows or a high proportion of heifers to cows. The most likely cause is poor feeding management but there could be others, such as disease, heat stress or simply poor reproductive practices.

8.6 Cow comfort; what it is and why it is so important

'Cow comfort' is a relatively new term that is being more commonly used to describe 'how a cow is feeling' about her surroundings. Some people just think of it as the type of bedding material available when the cow wants to lie down and rest. Others use it to describe the climatic environment, that is the zone of thermo-neutrality or the 'comfort zone', in which cows don't need to use their physiological processes (such as shivering or increased respiration rate or sweating) to maintain their body temperature. Others consider it is more related to the cow's ability to do what they want to do and when they want to do it. It is generally agreed that cow comfort should also be extended to its psychological as well as its physiological wellbeing. Therefore, the complete definition addresses climate stress, poorly designed and constructed housing, stock facilities and the potential behavioural stress arising from herd mates and stock people.

Whatever it is, the most important thing is that it does not upset the cow's appetite for feed or water. Feeding management is devoted to ensuring the quality and palatability of the forage and concentrates are the best possible, so the last thing farmers want is that cows do not want to eat what they are offered. In the tropics, the key to ensuring cow comfort is addressing heat stress issues.

8.6.1 Heat stress in milking cows

The comfort zone for milking Friesian cows is between 6 and 18°C. Between –5 and +5°C, appetite will be stimulated while at the upper level, above 27°C, appetite is depressed and both biological and economic efficiencies decline. Above 24°C, dry matter (DM) intake decreases by ~3% for every rise of 1.2°C.

As cattle have a limited ability to sweat, the main route of heat loss in cattle during hot weather is evaporative cooling from the respiratory tract, namely the nasal passages and lungs. Cattle then increase their breathing rate to increase movement of air over the moistened surface of the upper respiratory tract and mouth. However, if humidity levels are high, the effectiveness of this evaporative cooling is decreased and cattle may be unable to dissipate accumulated body heat.

Clinical signs of heat stress

The following signs can be used to assess the degree of heat stress:

- mild heat stress: drooling, increased respiration to 80–100 breaths/min
- moderate heat stress: drooling, respiration of 100–120 breaths/min and occasional open mouth panting
- severe heat stress: drooling, respiration rate greater than 120 breath/min and open mouth panting with tongue out. Cattle also have an agitated appearance, hunched stance and will often have their head down
- cattle can move from mild to severe heat stress very quickly, within 30 min to a few hours. Therefore, extra vigilance is required once mild heat stress is detected.

8.6.2 The Temperature Humidity Index and other measures of heat stress

Temperature Humidity Index

Heat stress occurs through a combination of high ambient temperature and humidity and other factors such as poor ventilation and high levels of internal body heat production.

The best single measure of heat stress is the Temperature Humidity Index (THI) as this combines temperature and relative humidity into a single comfort index. The relationship between temperature and humidity is presented graphically in Figure 8.4 and also in tabular format in Appendix 1. The higher the index, the greater the discomfort, which occurs at lower temperatures for higher humidities.

Its effects on cow performance are summarised in Tables 8.2 and 8.3. An air temperature of 30°C corresponds to a mild THI of only 74 at 25% humidity or 78

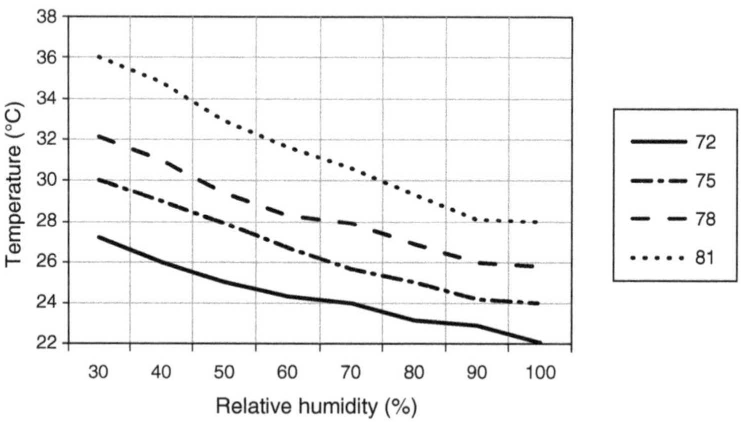

Figure 8.4: The effect of increasing relative humidity on the temperature to produce the same Temperature Humidity Index (THI).

at 50% humidity whereas it produces a severe heat stress (THI 82) when recorded at 75% humidity, a quite common occurrence in the humid tropics. For milking cows, an air temperature of 35°C equates to severe heat stresses (THI 80) at 25% and (THI 85) at 50% whereas it produces a very severe heat stress (THI 90) at 75% humidity.

Table 8.2. Effects of Temperature Humidity Index (THI) on dairy cow performance.

Comfort zone	THI	Stress	Comments
A	< 72	None	–
B	72–78	Mild	Dairy cows adjust by seeking shade, increasing respiration rate and dilution of blood vessels. Cow performance is adversely affected with reproduction more so than milk yield.
C	78–89	Severe	Both saliva production and respiration rates increase. Feed intakes decrease while water intakes increase. Milk production and reproduction are both reduced.
D	89–98	Very severe	Cows will become uncomfortable due to panting, high saliva drooling and high body temperatures. Milk production and reproduction will markedly decrease.
E	>98	Danger	Potential cow deaths can occur.

Table 8.3. Effects of Temperature Humidity Index (THI) on measures of cow heat stress.

Level of heat stress	THI	Respiration rate exceeds (no/min)	Body temperature exceeds (°C)
Zero	68	60	38.5
Mild to moderate	70	75	39.0
Moderate to severe	80	85	40.0
Severe	90	120–140	41.0

Some people argue that today's high-producing dairy cows begin to feel the effects of heat stress more quickly than was previously thought. In fact milk yield losses can become significant when the minimum THI on any given day reaches 65 or higher. Therefore, rather than set artificial cooling (say with fans and sprinklers) to begin when the THI reaches 72, they set the threshold to 65 to commence cooling at 65 to help cows deal more effectively with heat stress.

THI only quantifies the effects of air temperature and relative humidity, hence it neglects the other important aspect of heat stress, namely air movement. This was recently exemplified when the senior author visited a large dairy feedlot in the coastal lowlands of North Vietnam, only 40 m above sea level. In this 600-cow farm, the milking cows were kept in tunnel ventilation sheds whereas the young stock lived in non-tunnel ventilation sheds that depended entirely on internal fans and sprinklers. Data on temperature and humidity were collected every day at various times each day. The 2 p.m. THI data for the two types of sheds were plotted against the date. The average monthly THI inside the tunnel shed was the same as that inside the non-tunnel ventilated shed each time it was measured at 2 p.m. Mean monthly THI at 2 p.m. varied from 87 (in May) to 70 (in December).

In contrast to these observations, were others made in East Java in Indonesia at a large dairy feedlot that was located at 1200 m above sea level. At 9.30 a.m., the THI inside one of their tunnel ventilation sheds was 72 compared to 76 outside this shed. Being higher, the air was cooler inside compared to outside this shed, namely 24 v 28°C in May. In contrast to this, the Vietnamese shed was much hotter, with mean monthly 2 p.m. temperatures in summer (May to Aug) varying from 30 to 37°C. The impact of altitude on air temperature, which decreases by 6.5°C per 1000 m, meant that the outside air temperatures at the East Java feedlot would have been up to 10°C cooler than the Vietnamese feedlot, which in the morning would have increased the efficiency of the tunnel ventilation through considerably cooler air being forced through the input fans.

Therefore, people can quite rightly argue that THI may not be the best single indicator of heat stress. Another argument against THI is that since it only records the conditions at any one point in time, it cannot take into account any build-up in heat loads previous to the recording. One such Index has been developed for humans to do just this. The Accumulated Heat Stress Index accumulates the hourly heat stresses for 72 hr using time-weighted mathematical functions. Another index has been developed specifically for grazing dairy cows which takes radiative heat load into account. The Dairy Heat Load Index takes wind speed and direct solar radiation into account as well as temperature and humidity. This is not really relevant to intensive dairy systems based on feedlotting in sheds where radiative heat load is much less of an issue.

Heat Load Index

This is a new measure of heat stress developed specifically for feedlot beef cattle that are normally kept in dirt yards with minimal shade. It is based on a combination of:

- black globe temperature, which is a measure of the radiation heat load, taking into account both the ambient temperature and solar radiation
- relative humidity
- wind speed.

This index includes several adjustment factors such as genotype, coat colour, access to shade, water temperature in drinking troughs, and whether the animals are sick or healthy. Use of this index over time allows for the calculation of an accumulated heat load and the required heat loss during the night to maintain zero heat balance. However, developing such a Heat Load Index for Asian dairy cows normally maintained in sheds, is unlikely to provide an additional useful management tool.

8.6.3 Adverse effects of heat stress

For Friesians producing 20 kg/d, a THI above 78 leads to a decline in milk yield. There is also a decline in milk composition (milk fat and milk protein contents) but this occurs at 1–2°C higher than corresponding break points for milk yield.

With regard to reproduction, this declines before milk yield, namely, at THI of 72. Cows in early pregnancy (up to 3 weeks) can abort while cows in mid-pregnancy can have reduced birthweights. Cows are also more likely to have shortened and/or silent heats (less than 8 hr). Heat stress delays heat (hence submission rates) and, at the time of insemination or during the following 3 to 5 weeks, it can reduce conception rates and increase embryo mortality. By comparing conception rates between seasons (hot v cool or wet v dry), heat stress may be diagnosed as a problem if seasonal conception rates differ by more than 10–12%.

Severe heat stress will increase body (rectal and uterine) temperatures which in turn have a direct effect on reproductive performance. When rectal temperatures increase by 1.0°C within 12 hr after insemination, pregnancy rates can drop by 16%. An increase in uterine temperature of 0.5°C on the day of, or the day after, insemination can decrease conception rates by 13% and 7% respectively.

Cows are particularly vulnerable at temperatures above 30°C or, above 25°C with high humidity. Cows producing more than 15 kg/d of milk are more susceptible to heat stress due to their higher metabolic heat load. Zebu cows are less susceptible than Friesians because of their dense flat coat and higher density of sweat glands, however, exactly how less susceptible they are has not been documented. When planning strategies to minimise heat stress, it is then important to give priority to non-pregnant cows, usually in early lactation.

However, recent studies with transition cows (non-lactating and heavily pregnant) have shown that even though heat stress may not greatly reduce their voluntary feed intakes, it reduced subsequent milk yields by up to 10 kg/cow/day. This was attributed to a reduced inability of these cows to partition their body tissues towards milk synthesis during early lactation. Furthermore, the calves born from these heat stressed cows had lower birthweights, reduced growth of mammary cells during early udder development, which led to them also producing less milk, by up to 5 kg/cow/day, in their future lactations.

Recent surveys of the seasonality, particularly the adverse effects of high THI, on dairy cow performance in the United States, have shown marked adverse seasonal impacts on calf birthweight, yields of liquid milk and milk protein as well as conception rates and early foetal losses in pregnant cows. The study also highlighted the threshold THI for death rates of mature Friesian cows. Above the daily minimum THI of 70 and/or the daily maximum THI value of 80, there was a break point (indicating a sudden increase) in the incidence of Friesian cows dying, presumably from the additive adverse effects of heat stress. It is of interest that these values have decreased in recent years because of the improved genetic merit of dairy cows worldwide to produce more milk and the increasing number of hours per day that they are likely to experience heat stress (Young 2015). Some people might argue that this is the result of recent evidence on global warming.

In addition to reducing appetite, and hence milk yields, heat stress in milking cows has been shown to decrease lying times, reduce the proportion of saliva reaching the rumen and so affect its pH (thus making the cows more susceptible to subclinical rumen acidosis) and also impair reproductive success (both greater pregnancy losses from 21 to 30 days of gestation and more inseminations per successful conception), while increasing the time spent drinking, the incidence of lameness, hence cull rates and even cow mortality rates. It also changes the type of energy metabolism in the cow by reducing body fat mobilisation and increasing muscle breakdown, which would reduce the cow's ability to stimulate milk production in early lactation through repartitioning energy from body reserves to the mammary gland.

More recent evidence indicates that heat stress can have long-term detrimental effects during the cows' dry period. Heat stressed dry cows have depressed appetites, they lose more body condition and have a reduced immune function to combat disease/infections. Their milk yields are also adversely affected during their next lactation. In addition to this, the calves born to these heat stressed cows have lower birthweights, growth rates and reduced immunity, through lower absorption rates of maternal immunoglobulins from the colostrum to their blood stream, which has led to higher pre-pubertal culling rates. This is associated with adverse long-term effects of lower milk production and more inseminations per successful conception as mature cows. These recent findings clearly indicate the importance of paying much closer attention to heat abatement procedures in pregnant, dry cows (Young 2015).

Mating heifers can respond to heat abatement following just 4 h cooling using fans and sprinklers. This has been shown to increase their pregnancy rates from 23% to 57%. Herd bulls are also adversely affected by heat stress through reduced spermatozoa counts; this occurs 6 to 8 weeks following the heat stress period.

Adverse effects of heat stress are delayed by several days. The effect of mean THI two days earlier has the greatest influence on milk yield, while the effect of mean temperature two days earlier has the greatest influence on feed intake. Another good 'rule of thumb' when assessing heat stress for dairy cattle is that air temperature (in °C) added to humidity (in %) should ideally not exceed 90.

Improvements in milk yields of up to 3 to 5 kg are possible through effective cooling strategies. For example, in the East Java tunnel ventilation shed mentioned previously in Section 8.6.2, which housed Friesian cows each producing over 30 kg/day, it was estimated that this additional cow comfort produced an additional 3 kg milk/cow/day. For the record, as a result of this dramatic improvement in cow milk yields, the management of this Indonesian dairy feedlot has since installed tunnel ventilation in every dairy shed on the farm.

The high humidities associated with high daytime temperatures in the humid tropics reduce the effectiveness of using water to cool the air, thus negating the beneficial effects of evaporative cooling. Increasing respiration rates are the cow's natural way to increase evaporation from the alveoli tissues just as actually wetting the cow's skin provides the best opportunity to reduce blood, hence body temperatures, through cutaneous evaporation. Obviously increasing air movement over the skin, using fans, is the best way to stimulate such cutaneous evaporation. Therefore, initially using fans, followed by complete saturation of the cow's coat, using a computer/mechanically controlled cycle is the most effective artificial method of dissipating heat stress.

From the above evidence, heat stress clearly adversely affects many aspects of the performance of our modern day, potentially high yielding, milking cows. Therefore, when developing dairy development programs in the humid tropics and setting target milk yields, consideration must be given to more than just the type of milking cow and its feeding management. Farm management practices and physical shed designs to minimise the adverse effects of heat stress, are of equivalent, if not greater importance, in achieving such target milk and fertility measures of cow performance.

Cattle breed can play an important role in heat stress management. For example, Jersey, Brown Swiss and Friesian x Jersey stock are much more heat tolerant that purebred Friesians, and with good management, can also produce high levels of milk production. So they should be considered more for use in tropical dairy farming. *Bos indicus* cattle and buffalo are also quite heat tolerant.

8.6.4 Overcoming climatic stress

Observing the behaviour of cows is important in deciding when to modify management. If respiration rates reach 70/min, milk yield and reproduction may be compromised; this corresponds to 39°C body temperature, in contrast to a normal body temperature of 38.5°C. It is likely that higher yielding cows have faster respiration rates, because of the extra body heat production associated with higher feed intakes and milk yields. Improvements in milk yields of up to 3 to 5 kg/d are possible through effective cooling strategies.

Management strategies to overcome heat stress include:

- use more heat tolerant breeds of dairy cattle
- sheds should be designed to maximise natural air ventilation through its aspect on the farm, height and slope of the roof and having open sides
- cows can be cooled with hoses, sprinkler systems and cooling fans
- outside yards will provide for better overnight cooling as well as assisting in heat detection

- as forages produce more internal heat than concentrates, the bulk of forages should be offered in the cool of the evening
- drinking water should be continually available to milking cows
- Appendix 6 presents plans for cooling systems for a free stall shed.

8.7 Other measures of herd management

8.7.1 Proportion of cows milking of those that have calved

One good measure of the performance of the milking herd is the proportion of cows actually producing milk. For herds with a 12-month calving interval, lactation length should be 300 days (for a 65 days dry period), so lactation length would be the calving interval less 65 days, meaning that 82% of the cows are milking at any one time with 100% calving rate. However, in most year-round calving systems, less than 75% of the adult cows are milking. The longer the dry period, the fewer cows milking at any one time. The number of cows milking as a percentage of the total cow herd is influenced by several factors, the most important being lactation length, inter-calving interval and calving rate. It is assumed that cows with a 12-month inter-calving interval were dried off 65 days before calving. It also assumes no cows were culled for poor fertility or production and there were no mortalities among the milking herd.

Another way to demonstrate the importance of having as many of the adult cows milking as possible is to calculate the % days any one cow is milking which is related to the herd's inter-calving interval, the length of the dry period, hence the average lactation length. This is essentially the same as calculating the % adult cows milking for 100% calving rate.

Suggested KPIs for percentage of cows milking of those that have calved are:

- 74%; excellent
- 60–73%; acceptable
- 50–59%; below average
- 40–49%; not good.

8.7.2 Proportion of cows milking in the total dairy herd

Another useful measure of the proportion of productive cows is the size of the milking herd as a percentage of the total dairy herd, which includes the milk-fed and weaned replacement dairy heifers, breeding bulls (if any), dry cows and milking cows. As well as lactation length, inter-calving interval and calving rate, other important factors are heifer wastage (a combination of pre-weaning calf mortality and losses between weaning and second calving), age at first calving, culling of cows for poor performance and mortalities among the milking herd.

A series of assumptions had to be made on other key variables, namely a lactation length of 300 days, calving rate of 90%, half of the calves born were heifers, 10% of these heifers were sold before calving and the annual culling rate for the milking herd was 35%.

Suggested KPIs for percentage of cows milking in the total dairy herd are:

- 48%; excellent
- 40–47%; acceptable
- 35–39%; below average
- 30–34%; not good.

8.7.3 Average days of lactation for the milking herd

Although the target calving interval is 12 months, this is rarely achieved even in very well-managed herds in temperate developed countries as this requires cows conceiving within 85 days post-calving. This also means that the target 300-day lactation is also rarely achieved. If it is less than this, it is highly likely that the cow dried herself off early because of poor feeding management. It is usually greater than 300 days because the farmer plans to milk her until just 60 days before her calving down again.

Assuming a 300-day lactation length, the average days of lactation for any large milking herd that calves year-round should be 150 days. The greater the average days of lactation, the greater the number of days between calving down and a successful conception, that is, the greater the number of days open. Therefore, the average days of lactation can be a useful measure of the success of the herd's reproductive management.

8.7.4 Wet versus dry milking cows

There are various ways to categorise milking cows on any dairy farm, tropical or temperate. Variables such as daily milk yield (for example, less than 8, 8 to 12, 12 to 16, more than 16 kg/cow/day) and stage of lactation (early, mid, late or non-lactating) are the most common categories used on tropical farms, large or small. There are other ways of categorising cows, which are just as easy and even more informative that are worthy of discussing. This section describes two even simpler categories and what they can tell us about the feeding and herd management on any farm. These are lactation and pregnancy status (Moran 2015).

Adult cows are either lactating (wet) or non-lactating (dry). In the process of their full lactation they are either non-pregnant or pregnant. Pregnancy status is best determined through pregnancy diagnosis (that is an internal examination of the uterus by an experienced technician or veterinarian) but can also be ascertained by 'return to service' (that is whether the cow cycles in ~21 days since she was last inseminated or serviced by a bull).

Each cow would be given one of four possible statuses as follows:

1. Wet and non-pregnant (W/NP) when the cow calves down and before she conceives, usually some time during early lactation.
2. Wet and pregnant (W/P), following conception and up to when the cow is dried off (either naturally or through management).
3. Dry and pregnant (D/P), between drying off and calving down; this determines the minimal length of the dry period.
4. Dry and non-pregnant (D/NP), which should not occur but unfortunately often does.

Using a range of typical scenarios on any tropical dairy farm, the following tables provide data on the minimal length of the dry period and the proportion of cows (including first calf heifers) in each category. The D/NP category is not included in this table because on any well-managed farm, all cows should successfully conceive some time during their lactation. It is assumed that the gestation period is 280 days in length. The calving interval can be as low as 360 days, but is typically more like 400 days on most well-managed farms. The following tables are for year-round calving herds where the data represents the proportion of the herd in any one status on any one day during the year.

These scenarios for herd averages listed in Tables 8.4 and 8.5 are for:

- days from calving to conception; which is assumed to occur on average either at 90, 120 or 150 days into lactation
- lactation length: this is assumed to range from 240 days to 330 days in monthly steps (of 30 days for each month).

Table 8.4. The influence of days from calving to conception and lactation length on the calving interval and on the length of the dry period.

	Calving to conception (days)		
	90	120	150
	Calving interval (days)		
Lactation length (days)	370	400	430
330	40	70	100
300	70	100	130
270	100	130	160
240	130	160	190

It is recommended that cows are dried off at least 60 days before they are due to next calve down to allow the udder to fully recuperate in preparation for the next lactation. Therefore, the dry period in cows conceiving within 90 days of calving and milking for 330 days would be too short, hence they should ideally be dried off after 310 days milking. From Table 8.4, with calving intervals ranging from 370 to 430 days (or 12.3 to 14.3 months), the associated dry periods range from 40 to 190 days. Clearly, the earlier cows conceive after calving and greater proportion of their calving interval that the cows are actually milking for, the greater the income generated through milk production.

Table 8.5. The influence of days from calving to conception and lactation length on the percentage of milking cows in the herd in one of three cow categories, namely wet/non-pregnant (W/NP), wet/pregnant (W/P) or dry/pregnant (D/P).

		Calving to conception (days)		
Lactation length (days)	**Status (% herd)**	**90**	**120**	**150**
330	W/NP	24	30	35
	W/P	65	52	42
	D/P	11	18	23
300	W/NP	24	30	35
	W/P	57	45	35
	D/P	19	25	30
270	W/NP	24	30	35
	W/P	49	37	28
	D/P	27	33	37
240	W/NP	24	30	35
	W/P	41	30	21
	D/P	35	40	44

There are no additional nutrient costs of pregnancy to milking cows if they are non-pregnant or are in their first five months (150 days) of pregnancy. So the occurrence of pregnancy is unlikely to adversely impact of the cow's milk yields and/or feed efficiency until the last few months of pregnancy, which generally occurs during their dry period.

The important numbers in Table 8.5 are then the percentages of cows in the herd that are dry and pregnant because this is the period (once lactation ceases) they must go through to grow the calf *in utero* and when income generation falls to zero. With cows conceiving earlier after calving and milking for longer, these cows can number less than 20% of the herd. However, if conceptions are delayed and the cows are suffering from short lactations, this number can exceed 40%.

In both these cases, namely delayed conception and early drying off, close attention to feeding management is of the utmost importance to address these problems. This is essential first, so cows can minimise the period of their weight loss, to allow the post-calving hormonal cycle to initiate the oestrous cycles that are essential for successful ovulations and consequential conceptions. Second, it is necessary to provide the necessary supplies of feed nutrients within the udder to produce sufficient long-term quantities of milk precursors to minimise any likelihood of nutritionally induced short lactations. This is particularly relevant to many tropical SHD systems as the genetic merit of their cows is rapidly improving through the use of improved dairy genes either through imported semen and/or live animals. Unfortunately, associated with this is an increasing occurrence of repeat breeding as well as short lactations arising from inadequate supplies of feed nutrients, particularly energy and protein.

Dealing with dry/non-pregnant cows

Table 8.5 does not deal with the fourth category in milking herds, namely the D/NP cows. These cows will be non-productive for many months and will cost money every day for at least the next 280 days. If their condition is nutritionally induced, without any change in feeding management, it is likely to be a lot longer than 280 days before they generate any milk income. If the condition is due to animal health issues, then that urgently needs to be diagnosed and treated. A decision will also have to be made as to whether such animals should remain in the herd or be sold as non-productive cull cows.

In summary, dairy cows within any herd of any size can be easily identified as W/NP, W/P, D/P or D/NP to provide an additional tool with which to approach improved feeding and herd management. All it requires is a good set of eyes, a calculator and a notebook. Over time, as we collect additional data, we will be able to provide more definite herd management guidelines, such as threshold percentages of D/P and D/NP (ideally zero) when immediate action is required.

In addition, with further data collection it will be possible to put monetary values on these percentages for the likely financial returns arising from investing in the required improved feeding and herd management.

8.7.5 Persistency of milk yields

The term persistency relates to the ability of each cow to maintain a small steady decline in milk yield as their lactation progresses and is defined as the % monthly decline from peak milk for each month after the peak yield. The higher this number, the faster the rate of decline, hence the less milk produced over a full lactation. In Asia, lactation persistency of less than 8% per month is achievable on very well-managed farms, but more realistic levels are 8 to 12% per month. Assuming peak yields occur during the second month of lactation, it is possible to calculate full lactation milk yields for any cow with a given peak milk yield and persistency. Therefore, over a 300 day lactation:

- cows with a peak milk yield of 15 L/d and 8% persistency produce 3054 L over their full lactation or 10.2 L/d on average
- cows peaking at 20 L/d with an 8% persistency produce 4122 L total milk yield or 13.7 L/d average
- cows peaking at 25 L/d with an 8% persistency produce 5115 L total milk yield or 17.1 L/d on average.

Table 8.6. Full lactation milk yields (and average daily yields) in cows with varying peak milk yields and milking persistency.

Peak milk yield (L/d)	Persistency (%/month)	Monthly milk decline (L/d)	Full lactation yield (L)	Average milk yield (L/d)
15	4	0.6	3702	12.3
	8	1.2	3054	10.2
	12	1.8	2406	8.0
20	4	0.8	4986	16.6
	8	1.6	4122	13.7
	12	2.4	3258	10.9
25	4	1.0	6195	20.7
	8	2.0	5115	17.1
	12	3.0	4035	13.5
30	4	1.2	7479	24.9
	8	2.4	6183	20.6
	12	3.6	4887	16.3
35	4	1.4	8688	29.0
	8	2.8	7176	23.9
	12	4.2	5664	18.9
40	4	1.6	9897	33.0
	8	3.2	8169	27.2
	12	4.8	6441	21.5

Table 8.6 presents the full range of milk production data, for cows with peak milk yields varying from 15 to 40 L/cow/d and with monthly milk persistency varying from 4 to 12%/month. Full lactation yields then range from as low as 2400, as expected in traditionally managed lowland smallholder farms, to as high as 9900 L/cow, as achievable in well-managed dairy feedlots in the Asian highlands. Corresponding average milk yields range from 8 to 33 L/cow/d. Clearly feeding and herd management should be directed towards ensuring a 'flat lactation curve' throughout the entire lactation.

A very high rate of decline, indicating a rapid drop off in milk yield post peak, is often indicative of poor feeding management during early and mid lactation which often, particularly in high quality dairy cows, leads to rapid weight losses and delays in the first post-calving oestrus, thus resulting in reduced herd fertility. Therefore, for these cows in particular, feeding management must be directed towards supplying adequate nutrients, particular energy, in early lactation to achieve high peak yields, and in mid lactation to maintain milk yields with low persistency values.

The reason why persistency is an important measure to monitor is that when high genetic merit cows are underfed, they have been selected to utilise their body reserves to provide the nutrients in the mammary tissue to maintain milk yields. In other words, they have been bred to preferentially partition dietary nutrients from their fat reserves (mainly abdominal) to the udder. In so doing, they must lose bodyweight and this reduces the hormonal balance in the body necessary to allow ovulation and so allow the cow to show oestrus. This is called lactation anoestrus. Therefore, such cows will delay oestrus, hence have more days open. Such cows then require preferential treatment to bring them back into oestrus and need to be rapidly identified. Milk persistency can then be used to locate such cows that are more likely to have poor reproductive performance.

8.8 Milking management

The key to producing high quality milk is good hygiene and its rapid cooling following milking. Hot water is essential in any milking shed to ensure all milking equipment can be cleaned and sanitised.

Milking machines

As it is assumed that 500 cows are too many to milk by hand, all sheds will be using milking machines. The principles of maintaining milking machines are the same whether they are mini bucket milkers with one or two clusters or larger machines in a milking shed with several milk clusters. Pulsators should run at 50 to 60 pulsations/min while vacuum pumps should run at 40 to 50 kPa. All milking equipment must be routinely serviced.

One of the most common causes of mastitis outbreaks is inappropriate milking machine function. The need for routine milking machine maintenance cannot be over-stressed.

Rubber ware

From the functional aspect of a milking machine the most important piece of equipment on the machine is the liner or inflation. The inflations are the only part of the machine that comes in contact with the cow. From the day they are fitted to the milking machine they begin to deteriorate and lose their flexibility. A good rule of thumb is to replace the inflations after 2500 milkings, using the following simple formula:

$$\text{Replacement age (days)} = \frac{2500 \times \text{number of claws}}{\text{herd size x milkings/day}}$$

Apart from functional aspects of the inflation, the other concerns are the milk quality problems caused by deteriorated rubber ware. Long milk rubbers should also be replaced regularly (usually every 9–12 months).

Milking routine

Operators must be well trained and understand the need for routine and effective milking routine and management. They should wear disposable gloves and preparation should include cleaning if the teats are dirty. This is best managed using single wipes for each cow. This can follow pre-dipping, using a separate section of the wipe to clean each teat, or use them dry on each teat. Prior to attaching the milk clusters, one to two squirts of milk should be stripped out of each teat onto the floor, or a strip cup for closer examination. The operator must also be aware of any swelling and heat in the udder. Any sign of mastitis needs to be recorded and reported and possibly treated, according to predetermined mastitis management program SOPs.

This preparation will also provide the necessary pre-stimulation for letdown to occur. Teat wipes can be new paper towels or re-usable cloths that are thoroughly washed and dried completely following each milking.

If teats are not cleaned, they need to be gently massaged by hand, stripped and then allowed to rest for at least 30 s before clusters are placed on teats, to allow for milk letdown to occur.

Cups need to be removed when milk flow has almost ceased. Massaging the last of the milk out of the udder is not recommended, as this may create vacuum in the udder and reverse airflow into the teat and udder after cluster removal. Since over-milking will cause teat-end damage and can lead to mastitis, the small amount of milk remaining is not an issue.

Teats need to be effectively dipped or sprayed with a recommended and freshly prepared post-milking teat dip. Cows then need to walk on clean laneways and kept standing for 30 min to one hour.

If teats and udders are excessively dirty, running water may be needed to clean them properly. However, the teats must be dried, with a new paper towel, before 'cups-on'. In this situation, housing and bedding, as well as laneway issues need to be addressed in order to ensure clean and dry teats and udders at all times. It must be stressed that poor environmental conditions will ultimately lead to mastitis. Furthermore, fly control is also important.

9

Developing your observation skills

This chapter discusses the importance of observational skills and practical ways to improve them, to achieve efficient animal husbandry skills, as well as improved animal health, welfare and productivity outcomes.

9.1 What to look for

There are many ways of assessing how well a dairy farm is being managed, even before you interview the farmer. This chapter lists many of these, highlighting the importance of developing our observation skills so we can independently more quickly determine how well the farm has been planned, constructed and run on a day-to-day basis. This can be broken down to several steps:

- the farm's physical facilities
- the general level of tidiness and cleanliness of the shed and farm infrastructure
- the physical state and wellbeing of the livestock
- animal management practices.

9.1.1 Observing the farm's physical facilities

- Location with respect to the nearest bulk milk-handling facilities.
- Year-round access by road.
- Proximity of cowsheds to forage sources.
- Daylong shade for stock in sheds.
- Shed design to aid natural ventilation (roof and walls).

- Air temperature, humidity, breeze, dust and level of ammonia (close to bedding).
- No cobwebs in roof area.
- Sufficient space for all stock to rest comfortably.
- Sufficient space to prepare and feed out formulated ration.
- Sufficient yards to house different categories of livestock.
- Adequate stock handling facilities, such as race and head bale.
- Shed design to minimise obstructions and hazards for stock.
- Ideally the shed has loose housing and/or free stalls.
- Ideally there should be an outside yard where stock can rest during the cooler evening.
- Cow laneways are non-slip but also do not cause sore feet and lameness.
- Good drainage inside shed leading to an outside effluent pond.
- Adequate area and bedding for stock to rest in.
- Evidence of adequate clean drinking water supplies for all stock.
- Evidence of hot water for cleaning milk-harvesting equipment.
- The 'calf kitchen' is clean and well organised.
- Office area for writing and storage of farm records and other paperwork.
- Relaxing/lunch area for farm staff.
- Some effective biosecurity measures are in place.
- Easy access to any farm records, for both production data and financial management.
- Other value-adding operations, such as biogas.

9.1.2 Assessing tidiness and cleanliness around farm and inside sheds

- Clean stalls, laneways, alleyways and feed bunks.
- Feed storage area clean and tidy with effective vermin control.
- No spilt feed around the farm where flies and vermin can breed.
- No muddy tracks where tractors drive.
- Farm equipment and machinery clean and maintained.
- Minimal rubbish lying around farm outside cowsheds.
- Minimal rubbish lying around inside cowsheds, feed storage and milk-harvesting areas.
- Lockable veterinary cupboard, maybe with small refrigerator for storing some drugs.
- Clean area for stock to lie down (may have just been cleaned specifically for visit).
- Good stock cleanliness; look particularly for any dried manure on the back and side of the udder, the thighs and hips and the legs and hooves.
- Drinking troughs are clean.
- Feed troughs are not full of old rejected forages and other feeds.
- The shed effluent system is organised to ensure all the effluent ends up in a pond, rather than just lying around at the lowest point of the farm at the end of the cowshed.

9.1.3 Assessing the physical state and wellbeing of stock

Many of the following 'cow signals' have quantifiable scoring systems which are fully described by Moran and Doyle (2015).

- Cattle lying in stalls and not in alleyways.
- All stock are bright and alert.
- Body condition; muscle and fat cover between the pin bones and the tail head, over the hips and covering the lumbar vertebrae.
- Locomotion and lameness score; by observing back posture, head and limb position and behaviour when walking.
- Hoof score; will need to be able to restrict the animal to lift up legs to inspect hooves.
- Leg scoring; assesses the stance of the hind legs.
- Rumen scoring; this quantifies the intake of feed and its rate of passage through the gut over the previous few hours.
- Rumination; over 60% of the cows ruminating and lying down.
- Manure scoring: pay close attention to consistency or moisture content and the level of digestion as evident from the physical state of the manure.
- Teat scoring; assess both the teat ends and the skin around the teats.
- Respiration rate to assess the degree of heat stress: this is quantified by standing behind the cow, counting the number of breaths over 10 s and multiplying this number by six. Respiration rates exceeding 75 to 80 per minute indicate some degree of heat stress; it is severe if it exceeds say 120 breaths per minute.

9.1.4 Assessing animal management practices

It is important that stockmen carry out animal handling practices in ways that are effective, humane and result in as little negative impact on the animals as possible, and that they exhibit proper care and attention to the needs of the animals under their care. This includes:

- appropriate milking management and milking hygiene
- appropriate mastitis management practices
- animals are moved at their own pace and not forced, and are handled with due care
- calves are dehorned appropriately
- cow bedding is maintained
- parasite control vaccination is administered appropriately
- AI is conducted appropriately
- obstetrical interventions are administered appropriately
- cow, bull, heifer and calf feeding and watering regimes and practices are implemented appropriately
- shed cleaning is implemented appropriately

- sick cattle are diagnosed and managed appropriately
- health management plans are implemented appropriately, including lameness prevention and management
- abnormalities and events (such as calving, oestrus, disease, cows panting, animals not eating, water troughs empty etc.) are recorded, reported and acted on as soon as possible
- diseases are treated and managed – such as mastitis and lameness.

9.2 Don't just look; observe your stock

As Hulsen (2011, 2013) states in his practical manuals on cow signals, the steps involved in using cow behaviour in farm management decision-making are:

- What do I see? This involves careful observations that should be described objectively and precisely. Is what I see normal or abnormal?
- How has this come about? Why is this happening? What are the causes?
- What does this mean? What should I do? What are the practical solutions to any obvious problems?
- In essence, it is *look*, *think*, then *do*.

Another reason for looking at cow signals more purposefully is to overcome the danger of 'farm blindness', that is thinking that what you see every day around the farm is normal. Farmers should always ask, 'What is normal on my farm, and is it the same for all farms?' It is essential then to make a point of including specific observations in your daily routine, discuss such matters with your farming colleagues and service providers and also to visit other farms to note how these specific observations may differ from your farm.

Astute observers should take into account:

- Understanding what is normal in anatomy/shape, posture, behaviour, movement, reactions so you are able to detect what is abnormal.
- Not just looking, but looking and observing. For example, you must notice all the signals that stock send out because they can all provide important information about cow wellbeing and comfort.
- Focused observation. You must look for things to evaluate. Ask the question 'Is everything as it should be or might this situation pose a potential risk?'
- Open-minded observations. Look at things as if for the very first time and forget any excuses and preconceived ideas.
- Compare your current observations with some form of standard that indicates whether additional action needs to be taken. You will then need to develop some standards for your particular situation.
- Observe from large to small, from many to few and from far to near. For example:

- Is the herd uniform in size, coat condition, cleanliness, body condition, abdominal fill? If not, why not? Is any non-uniformity important?
- How are the animals distributed within the building (particular cubicles, throughout the laneways, on the periphery of the buildings) and if it is non-uniform, why? Is it important?
- How many cows are lying down in the cubicles? While resting, is it close to 85%?
- How many cows show abnormal posture while walking? Is it sufficient to cause concern about lameness?
- After looking at the big picture, zero in on certain areas of the shed or on specific cows worthy of additional observation.

- Ensure there is an observation routine in which every animal receives some attention, with cows (milking, dry and transition) observed at least three times daily and heifers and bulls at least twice daily.
- If possible, take into account cause and effect. For example, if there is a physical deformity: what might have caused it, such as swollen feet and only access to concrete floors? (See Figures 9.1 and 9.2.)
- In addition to your eyes, nose and hands, 'paper information', namely records, can greatly add to the observations. This highlights the value of good record

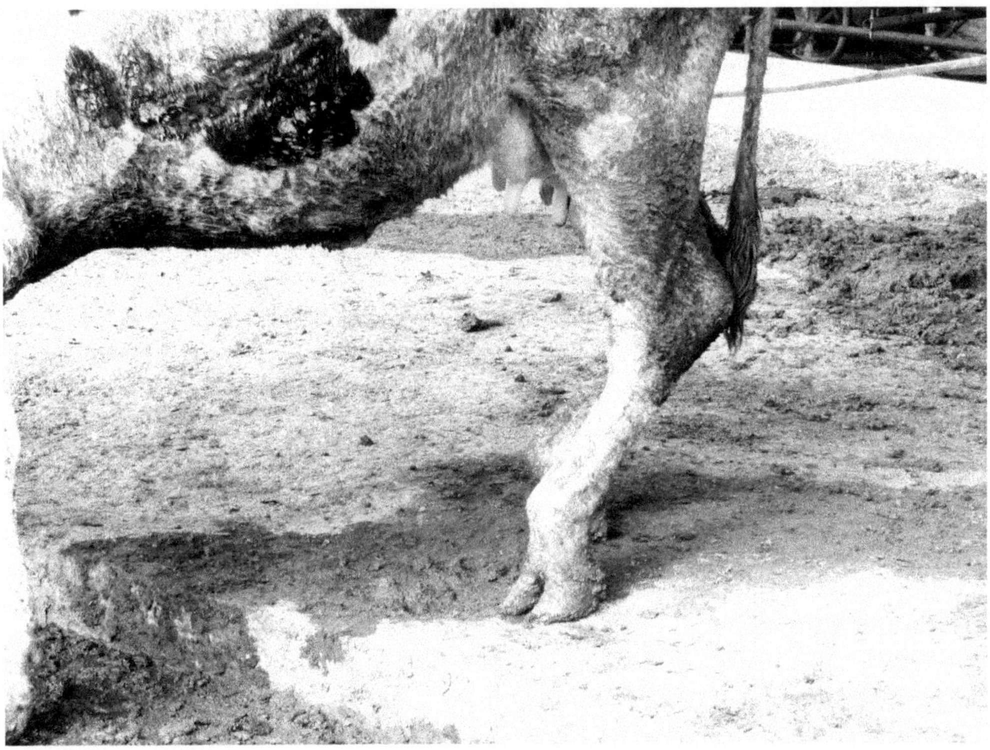

Figure 9.1: This cow is clearly suffering from a deformed hoof, which will adversely impact on her performance.

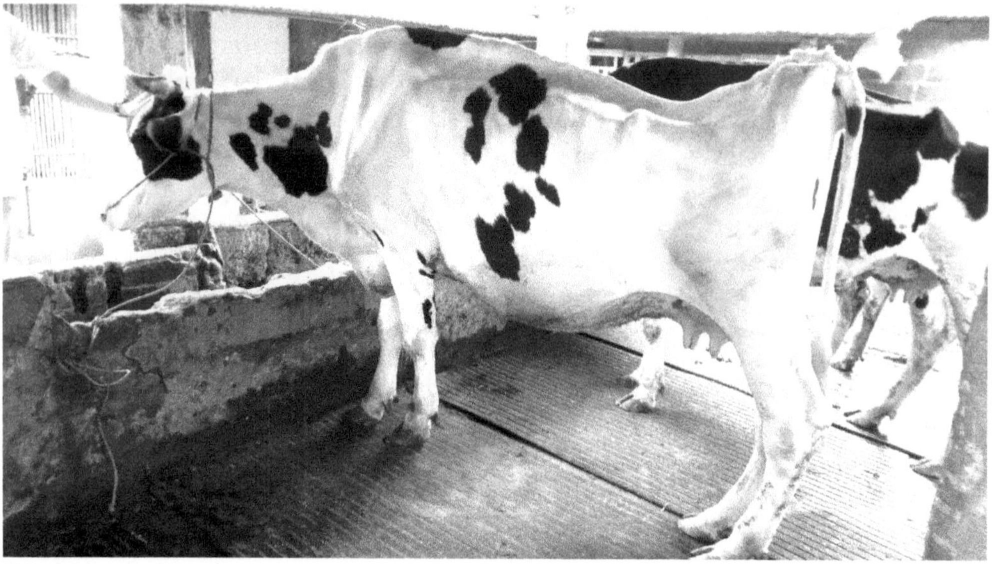

Figure 9.2: An old cow with permanent skeletal deformities as a result of a lifetime of tethering.

keeping that can be easily accessed to supplement the physical signals emitted by the cow.

- The shed layout should be designed for easy observation such as having gaps in feed barriers and perimeter fences and a centrally located cattle crush for ease of closer individual animal observation, if necessary.
- Record the key observations and any follow-up actions. Writing things down is recommended as it forces you to more clearly describe what you see and facilitates information exchange with farm staff and, if necessary, veterinarians or other professionals. For example, Figure 9.3 indicates a problem with a cow lying across two free stalls.

9.2.1 Signs of animal wellbeing

Assess whether the herd is uniform or if there are marked differences between animals. Pay attention to:

- Animal development: are heifers much smaller than the cows? If so, focus more attention on heifer rearing.
- Body condition: when more than 10% of the cows are too fat or too thin. This indicates a long-term imbalance between feed intake and utilisation. Focus on feeding regime, trough space, availability of feed during the day, hoof health, the way cows select their feed and dietary fibre content.
- Hair colour, coat shine and cleanliness: a glossy coat is a sign of a healthy animal. A dirty coat is always a bad sign, and can indicate things such as the

Figure 9.3: There is insufficient sand bedding in this free stall because cows can lie across two free stalls.

need to change the bedding, scouring from illness or poor ration formation. Hair missing on legs may indicate poor bedding.

- Abdominal and rumen fill: these indicate feed intake over the last 24–48 hr. Why did the cows eat less? Are we dealing with a risk group (see Section 9.4.2 below)? High yielding cows and those close to calving must reach their optimal feed intake as soon as possible.
- Abnormal posture and gait may indicate abdominal or leg pain. Which leg is painful? Is it more than one leg or one cow?
- Signs of boredom, for example constant licking or tongue extensions, head pressing, rocking etc. Possibly install back scratchers and other cow 'toys'.
- Other signs: are there common abnormalities in the herd? For example, consistently located lump on many cows' shoulders could indicate improperly constructed or installed feed yoke, or the cows have to reach too far into the feeding place to get their feed. Ulcers from lying indicate a need to improve or change bedding.

9.2.2 Logic of cow signals

There are generally three reasons for specific cow behaviour, namely:

- It satisfies a **need** and the cow wants something, for example she is eating, wants to lie down or is just plain curious.
- It is a reaction to a **stimulus**, for example she tries to avoid being physically hurt so she moves away from people or dominant cows or even jumps after touching an electric fence.
- It is due to a **physical urge** caused by pain, disease, hormones or she may be due to calve down.

These help answer the question, 'Why is the cow behaving in a certain way?' If you do not know whether such behaviour is normal, compare the cow in question with other cows on your own farm and then on another farm to assess the possible reasons in a completely different situation.

As genuine cow signals are repeated, for example kicking off a milk cluster only once may mean a single cow is overreacting, but does she do it at every milking and if so, why? If it is repeated many times, then there must be a common cause such as over-milking, vacuum level is too high, teat injuries, stray electric voltage or even severe fly irritation.

If a cow looks as if she intends to do something, makes an attempt to do it but then stops, there must be a reason why she did not follow through her intention. What were the circumstances or stimulus or stimuli that made her change her mind? Learning to recognise normal behaviour and then the things that might inhibit this provides valuable information about the underlying relations in a herd, housing or health of a cow.

An observation that defies logic can be an extremely valuable cow signal. Hulsen (2011) calls these unclassified notable observations (or UNO, 'you know'). At first glance these findings may appear to be insignificant, but on reflection and further consideration, the observation can become important. If it can be a potentially harmful UNO, it justifies an explanation. For example, if particular cow drinks water from dirty puddles, she must be thirsty, so providing additional or more accessible fresh drinking water would solve this problem. There may also be a mineral deficiency. When evaluating UNOs, use the same three steps listed above, namely:

- Describe exactly what you see.
- Ask yourself, or someone else, what things may have caused this?
- Determine what influence the signal has on comfort, health and production, and decide whether or not to take action.

Hulsen's (2011) book has many 'take home messages' and the key ones have been summarised in Table 9.1.

Table 9.1. 'Take home messages' about how cows and humans communicate with each other.

• Don't just look … observe.
• Like cows, use all your senses when observing; ears and nose as well as eyes.
• What do I see, why has this happened, and what does it mean?
• Observe from large to small, from many to few, from far to near. Then do this in reverse.
• If cow signals are genuine, they will be repeated.
• Why is a cow doing this? Does it satisfy a need, is it a reaction to a stimulus, is it due to a physical urge?
• Cows are herd animals, so tend to do things simultaneously, but they also form smaller groups within herds, generally based on social order.
• Cows use subtle signals to indicate their social 'class' or ranking (dominance or submissiveness).
• Conflicts often occur between cows of similar rankings.
• Cows tend to moo when they are on heat or are hungry.
• Each cow has a personal space and this varies in size between breeds and their feeling of security and trust.
• Cows feel secure when they know they have plenty of feed, they have escape routes and know the person.
• Cows feel insecure on slippery floors, when very lame and when in presence of unpredictable personnel.
• The presence of cobwebs in the cowshed is indicative of low air movement, hence poor ventilation.
• If more than 10% of the resting cows are standing, stall comfort needs to be improved.
• When cows lie down, up to 30% more blood circulates through the udder.
• Ration formulations rarely correspond to what cows actually consume. So they only act as a starting point and need to be verified and modified on the farm.
• The body language of the cow is the best management adviser. If you can read the cow, you know what to do.
• Cows will tell you what they want.

9.2.3 Indicator animals and locations

Indicator animals are those that belong to certain groups of stock on the farm that are at greater risk than others. They are often the first to send out signals indicating something is wrong. Observing abnormal behaviour from members of these high-risk groups can provide advance notice of a problem. For example, high-yielding milkers will be the first to show up a problem in the formulation of the ration, through unexpected drops in milk yield.

Indicator animals can also be used to monitor the likelihood of a potential problem occurring on the farm, such as a shortage of forage might first become apparent with the changes in milk yields in heifers in the milking herd. Poor handling may first become apparent when previously bolder cows are slower to enter the dairy.

Risk locations on the farm can identify where stock are more likely to be injured, such as a long rough track where small stones can injure hooves, or the calf shed where sudden changes in weather can upset calf wellbeing.

There are times of greater risk for different stock groups on the farm. These can be related to:

- the season, such as the middle of the dry season when soil moisture levels are at their lowest
- a particular date, such as the first extremely hot day in summer
- stock age, such as at weaning time in milk-fed calves
- stage of lactation, such as when the first insemination is usually due.

Additional observations (and actions if necessary) and routine preventative measures can reduce stress during these times to limit any likely problems. Responding rapidly to such problems can prevent serious consequences. It is important to plan ahead to assess whether everything is as it should be and potential problems can be quickly detected and acted upon.

9.3 Signals of good and poor health

When assessing cow health, proceed from large to small. It is best to observe with an open mind rather than make judgements or excuses. Having someone else come and look with you can help you see more and draw better conclusions. Appendix 2 presents pictures of cows in good and poor health.

Signals of good cow heath include:

- The cow is alert and active, she does what she wants to do and is aware of her surroundings. Her eyes and ears are attentive and she is curious about noise and other stimuli.
- She has a glossy, smooth, clean coat without any blemishes. Cows that do not feel well soon lose the shine from their coats and their hair may stand on end.
- She has a good appetite and drinks well. Food intake is evident from the rumen and abdominal fill. If food intake is poor for a long time, the cow will lose weight and, eventually, her body condition. When she is not drinking enough or is losing excessive fluids, the eyes become sunken and the skin becomes tight.
- She walks and stands without any signs of pain or discomfort. When in pain or lame, a cow arches her back and is reluctant to move. Irritation in the pelvic area causes the cow to hold up her tail. If having difficulty walking, the cow first makes movements indicating she is about to walk, followed by obvious head movements when she starts walking.

- The cow is well cared for, with good housing. Cow signals such as overgrown hooves, mange and lice are indicators of poor care and should have been attended to long ago. Unclipped udders and backs, and dung caked on the cow's skin all suggest a lack of care.
- Appendix 2 also provides a summary of good cow and poor health and welfare.

Farmers often notice sick animals because they look or act slightly differently from other stock in the group. The earliest signs are subtle, requiring skill and experience as well as effort to recognise them. It is important to look specifically at animals in risk groups and at risk times.

Cows in negative energy balance have elevated levels of acetone or ketones in their blood, milk, urine and on their breath and show nervous signs. Some people can smell acetone, even when several metres from the cow.

A high body temperature is an early and clear sign of disease and is part of the immune response and inflammatory process. A sick cow that is not running a temperature may have a digestive disorder or could be in shock, which occurs when blood circulation is failing. In that case, the cow is cold to the touch especially her ears, lower limbs and udder. Taking her temperature should be the first step in any diagnosis.

When in pain, stock try and reduce the pressure on the sore part, take shallow and rapid breaths and are less aware of their environment. They also eat and drink less, showing signs of dehydration. They will often withdraw from the group and, in the shed, choose to lie in a stall. Lame cows are more easily startled because they are less able to get away and this becomes very obvious on slippery floors.

9.4 Cow signals at the feed trough

Cow nutrition focuses on achieving maximum dry matter intake and a healthy rumen. There are many factors contributing to rumen health. Nutritionists tend to focus on issues such as ratio of energy to protein and ensuring the ration has sufficient fibre and minerals, but astute farmers should consider all the factors that can influence a cow's eating behaviour.

Calculated rations rarely correspond exactly to what the cow actually consumes, because of natural variation and the need to make assumptions. Therefore, the ration calculation only acts as a starting point, which needs to be verified and possibly modified in the shed.

When evaluating nutrition, health and production, you need to look to the past as well as the present with the aim of achieving even better results in the future. Information from the past helps you to learn and to understand the current situation. It can be used as the basis for setting new goals such as: 'Next year I want to produce 50 L more milk per cow and also halve the feet problems.' The cow signals you notice can be used to evaluate the current situation.

Overcrowding at the feed bunk results in increased aggression between cows. This leads to hoof damage and lameness as less dominant cows try to avoid dominant animals by turning away from them, causing them to twist their rear feet on an abrasive surface (concrete). Increased aggressive interactions can lead to even more severe claw damage. The potential for laminitis also increases as cows may consume fewer, but larger, meals or even have reduced feed intakes and spend more time on concrete rather than lying in the stalls. Headlocks reduce this aggression and improve access to feed by socially subordinate cows during peak feeding periods, by offering some physical separation between adjacent cows.

9.4.1 What nutritional signals to look for

A high yielding cow giving 30 L/day of milk has a rumen volume of 150 to 200 L and each day, consumes ~22 kg of dry matter and passes ~35 kg dung.

When looking backwards, important cow signals include:

- changes in body condition scores
- annual and monthly production figures and milk records
- number of metabolic diseases, such as displaced abomasum, milk fever and ketosis
- total number of illnesses
- number of cows culled with the reason for culling
- fertility records.

When assessing the situation and making changes, the important cow signals (and what they indicate) are:

- rumen fill (feed intake and rate of passage)
- milk production today and yesterday (feed intake and energy to protein ratio)
- dung consistency (feed intake and digestion)
- selective feeding (well-mixed ration and palatability of ingredients)
- daily feed residues (should be 5 to 10% of feed offered)
- feed wastage (feed intake and selective feeding)
- chewing the cud or rumination (fibre)
- hoof health (locomotion score)
- heat stress.

Close investigation of the cow's body fill around the abdomen can provide a guide as to how much a cow has eaten today, this week and this month (Hulsen 2013).

- A cow that has eaten well has a good rumen fill and belly fill.
- A cow that has not eaten enough **today** has a reduced rumen fill. This is apparent from the depressed left flank, located at the back of the ribs under the rear lumbar vertebrae, just in front of the cow's hip.

- A cow that has not eaten well this **week** has a reduced belly fill. This is apparent from the depressed abdomen, located midway down the abdomen under the ribs.
- A cow that has not eaten well this **month** has a reduced body condition score.

9.4.2 Risk groups

On every farm there are groups of animals that are susceptible to shortcomings in the ration. These groups then require closer monitoring than others. They can also be used as a means of monitoring risks, as indicator animals. Separating risk groups from the rest of the herd reduces the risks. For example, freshly calved cows and heifers should be separated from the rest of the milking herd.

The four main groups of risk animals are:

- **Heifers**: Risks include inadequate feed intake, acidosis and hoof problems. Rations too high in energy or protein result in udder oedema and soft hooves. If the mineral ratio is wrong, this may also contribute to oedema. In addition, these animals may not eat enough forage compared to concentrate, a particular problem with insufficient space at the feed barrier. This can lead to laminitis, which increases the probability of hoof problems. Following calving, these heifers are often bullied causing stress, low immunity, lameness, reduced feed intake and loss of body condition.
- **Freshly calved cows**: Risks include milk fever, ketosis and fatty liver, metritis and mastitis. If these cows do not eat enough soon after calving, they need extra care and attention. Calving down in a dirty pen can lead to udder and uterine infections.
- **Cows in first two months of lactation**: Risks include inadequate energy intake, acidosis, ketosis and displaced abomasum. During the first 6 to 8 weeks of lactation, low forage and high concentrate intakes can lead to acidosis. The acidic rumen is poorly filled, does not contract and the contents are mushy. The dung shows the signs of poor feed digestion, smells acidic and alternates between thick and thin. Cows do not ruminate properly or for long enough so often discarded cud is seen in the pen.
- **Cows at end of lactation**: Risks include getting overfat and reducing concentrate intakes too quickly. Condition score, milk fat and protein levels should be monitored. Getting too fat is due to too much energy and not enough dietary protein or a ration that is too rich for the level of milk production.

10

Addressing the problems of stock welfare

This chapter presents an overview on dairy cow welfare and how it can be assessed from cow behaviour. It also briefly discusses the management of stock surplus to the dairy herd.

Stock welfare is a key issue for both smallholder and large-scale dairy farmers, probably more so with smallholders because many of these farmers have had limited exposure to dairy cows, calves and heifers, hence are less aware of what constitutes normal and abnormal behaviour. With large-scale operations, when large numbers of stock are forced to coexist in close proximity, any key issues arising from poor stock welfare can be magnified and lead to even greater problems and usually lower productivity.

Clearly understanding and interpreting cow behaviour is the key to recognising existing and potential problems of cow wellbeing, comfort, and hence, welfare. The reader is directed towards one of the senior author's books on this topic, called *Cow Talk: Understanding Dairy Cow Behaviour to Improve Their Welfare on Asian Farms* (Moran and Doyle 2015).

10.1 What do dairy cows really want?

Not every aspect of their surroundings is equally important to the cow. Bos *et al.* (2009) list just nine key factors in any dairy husbandry system that have important impacts on dairy cow welfare, these being:

1. At least one comfortable and spacious resting spot for every cow: cows like to rest together as a group. Rest is a necessity of life for the cow.
2. Good feed and sufficient drinking water: the feed must enable the cow to maintain her body functions and produce milk. It must contain sufficient dry matter, energy protein, fibre, minerals and vitamins to allow normal rumen digestion. Cows are selective when it comes to food so it must be tasty, varied and fresh, and not contaminated with manure or saliva.
3. Complete freedom of choice to move within the area and within the herd: cows want to make up their own minds. They want to be able to get out of the way of higher-ranked animals and in large herds, they like to split up into smaller groups. Sometimes a cow wants to get away from the group. Yet, she still wants to be able to see and hear her herd mates.
4. Calm and predictable handling by people, so she can move at her own pace: cows like orderly lives and prefer to know what to expect. Driving them and other unfriendly treatment will cause stress.
5. No negative stimuli such as leakage of electrical current and cow trainers (electric wires placed on top of free stalls). Negative stimuli will cause (chronic) stress that adversely influences both health and welfare.
6. No impact of obstacles while getting up, lying down and during resting, as well as being able to lie down at some distance from other cows. Cows like to lie down in the way they would while at pasture. They want their own personal space, but may still like to lie close together.
7. A comfortable climate: this is one with a THI below 71, which is equivalent to 27°C at 30% relative humidity or 24°C at 70% humidity. This avoids the stresses from being too hot.
8. Passage ways and feeding areas with a nonslip, dry and clean floor without sudden changes in the level, texture or shadows. If the floor is too smooth, cows may slip; if too rough, they may damage their feet. Uneven, wet or dirty floors are detrimental to leg and hoof health.
9. Sufficient light during the day (more than 200 lux, which is equivalent to general home lighting rather than a well-lit office): cows must be able to see their surroundings properly, so they can recognise their herd mates, explore their surroundings or play with their companions. In addition, light is important for fertility, which in turn is in the farmer's interest.

Providing all these prerequisites will ensure a happy, contented and productive herd.

10.2 Problems of confinement

Dairy stock imported from Western countries have almost invariably been reared under grazing conditions, hence have not usually been exposed to a continual shed environment, as is common on most smallholder and large-scale farms. Compared to grazing, confinement creates specific problems such as:

- restricting opportunity to seek comfort, for example, if cows are only provided with cement floors
- creating problems of high humidity, which can be more detrimental than high temperature
- limiting opportunity for exercise, hence the need for routine hoof trimming (see Figures 10.1 and 10.2)
- increasing exposure to infectious diseases
- other health issues, such as mastitis and uterine infections when hygiene is poor during milking and calving
- creating problems of heat detection for artificial insemination
- requiring greater efforts into sanitation
- magnifying problems of social dominance
- increasing capital investment.

It is essential that when designing housing systems, farmers keep the following key principles in mind:

- To achieve cow productivity, farmers must address the twin issues of cow comfort and health, where health is the freedom from infections, injuries, pain, stress and metabolic problems.

Figure 10.1: A cattle crush specifically designed for hoof trimming.

Figure 10.2: Hoof trimming, halfway through the operation.

- To achieve cow wellbeing, farmers must address the twin issues of housing and management, where management covers all the normal farm practices associated with dairy farming.
- Addressing cow comfort and housing, means that the farm's physical facilities must be optimised before putting cows in sheds.

Cows grazing in a paddock have access to forages virtually all the time and to concentrates at least twice per day, at milking. These feeding goals should also be the aim when cows are maintained in sheds. They should be encouraged to make at least 12 trips each day from their place of rest to the feed and water troughs. This will only occur if cows 'feel good', have healthy legs and feet, and the route is safe and comfortable.

10.3 The basic elements of dairy cattle welfare

Stock welfare in the dairy herd can be broken down to six basic elements, namely:

1. ***Facilities and equipment.*** These should be well designed, maintained and operated to ensure a high level of animal welfare, minimal stress or chance of injury.
2. ***Provision of feed and water.*** These should be provided with consideration of environment, age, body condition and stage of lactation that ensures their health, wellbeing and productivity.

3. ***Herd management.*** Dairy cattle should be routinely managed in social groups, handled to minimise stress, injuries and disease and to promote good health and welfare. Environmental conditions should also be such that the cow is comfortable.
4. ***Humane destruction.*** Weak, ill or injured cattle and calves should be identified and treated appropriately or humanely destroyed using approved methods.
5. ***Staff competency.*** All staff responsible for managing and handling dairy stock should be competent in their tasks and aware of their responsibility for good animal care.
6. ***Preparation, selection, sale and transport of stock.*** Stock should be selected and appropriately prepared for transport to ensure they are fit for the intended journey.

10.3.1 Other general principles of animal welfare

Whether housed in cubicles, straw yards or cowsheds, in order to maximise performance and ensure satisfactory standards of welfare, the accommodation must provide for the livestock's basic needs. As an absolute minimum, the housing must provide a safe, comfortable, clean, well drained and dry lying area together with shelter from adverse weather. It must allow the animal to move around freely without risk of injury.

Cows at pasture choose to lie down for 12 to 14 hr each day, so a similar target should be achieved with stock in sheds. If cows spend less time lying down, they are likely to spend more time standing in loafing or feeding areas that can adversely affect rumen function and hoof health. Sheds should provide a minimum loafing area of 3 m^2 per cow with at least 3 m between rows of cubicles.

Most farmers believe that animal welfare is just good cattle husbandry. The major welfare issues facing dairy farming include:

- housing, movement and cow comfort
- lameness,
- castration
- dehorning
- branding
- transportation
- slaughter.

Lameness can be a major problem, both from the point of animal welfare and farmer profits. Farmers frequently only recognise 40 to 50% of lameness problems and these are often well advanced, making them difficult to treat. The cow's environment, both social and physical, and her ability to cope with it ultimately determine how bad lameness can become.

Early separation of calves from their dam is considered by some to be a welfare issue, even though it is practised for animal health reasons. That is, to improve the likelihood of calves receiving sufficient quality colostrum in their first 24 hr of life which will reduce the likelihood of future infection and disease.

It is important to maintain animals in a body condition that is appropriate for their stage of lactation, to house them in social groups in clean environments with adequate space to move around and rest comfortably, to maintain feed, medication and production records as well as compassionately handling any animals undergoing management practices such as dehorning or castrating and those that are injured or ill.

Most Western countries have strict codes of animal welfare covering transport, housing, handling and feeding. Farmers purchasing dairy stock from such countries may be expected to abide by the principles of that country's animal welfare codes. This is particularly the case with young stock.

10.4 Observing and learning from cow behaviour

There are five key comfort zones that cows interact with every day, particularly cows that are confined to sheds. Dairy managers and farm staff can learn a lot by observing cows, sometimes for quite lengthy periods, to interpret their level of comfort in each of these zones. Each is discussed below.

10.4.1 Stall area

The major goal of dairy farming is to produce as much high quality milk as possible and this is best achieved in a non-stressed cow with clean dry udder, that can rest for 10 to 12 hr/d and lie down, rise and move in and out of the stall area without fear of injury. The base of the stall and the bedding should be clean and dry and provide traction and cushioning for the cow. The cow should have lunge space to get up or lie down. The linear dimension of free stalls should be designed to fit the size of cows using them. When entering a cowshed, it is best to observe the cows before they are aware of your presence.

Cows need to lie down for at least 12 hr/d while high yielding cows would benefit from 14 hr rest/d. To evaluate if cows are comfortable or not, it is possible to use a **cow comfort index**. Whenever cows are not eating or being milked, at least 85% of the cows should be lying down in periods between milkings. A second index is useful in free stall sheds, namely the **stall standing index**. This is the proportion of cows standing that at least touch a free stall, such as standing with just two feet in the stall. Less than 20% of the cows should be in this group.

Look out for:

- *Stall use.* Is the free stall empty? Are cows standing with two feet in the stall (this is called perching)? Is the cow lying in the stall or the alleyway?

- *Cow's position when lying down.* Is she lying straight or diagonally? Are her feet outstretched? Is she curled up? Is she intruding into adjacent stalls?
- *Cow injuries.* What condition are her hocks in? Is there hair loss? Are there any swollen joints? Does she appear lame?
- *Cow cleanliness.* Are her feet, legs, flank or udder dirty?
- *Cows lying down or getting up.* Does she lunge straight ahead or to the side? Are there obstructions that she encounters? Does she bang herself abruptly?

10.4.2 Environment

Temperature, humidity and air quality all affect cow wellbeing and performance. To reduce heat stress in sheds:

- Ensure there is maximal natural ventilation, so reduce any barriers to air movement and check wall designs.
- Provide adequate water space and volume.
- Supply supplemental cooling with fans and sprinklers.

Behaviour to observe includes:

- *Air quality.* Does the air feel heavy or humid? Does it feel stale with little or no air movement? Is there a noticeable smell of ammonia or other gases close to the ground?
- *Cows panting.* Are cows panting or showing other signs of heat stress?
- *Cows bunching.* Cows may bunch when they feel air movement. They may also bunch to get out of direct sunlight.
- *Cows clustered around waterways.* Cows splash water to try and cool themselves. They may lie around waterers if water has been spilt.
- *Response to sunlight.* Cows avoid direct sunlight during hot weather. A wide free stall shed will keep direct sunlight out of stall areas.

10.4.3 Feed area

If cows are not comfortable while feeding, intakes and milk yields will suffer. The goal should be to provide unrestricted access for as long as possible. First, the feed should be on offer all the time, but this is very rare on most smallholder dairy (SHD) farms. Cows should have sufficient space to eat from a clean, durable, smooth surface and through an easily accessible feed barrier. Raising the feeding surface ~10 to 15 cm above where they are standing puts them at a more natural feeding position. Locating a pipe barrier ~120 cm above where the cow stands and 15 to 20 cm beyond where the cow is standing, provides extra reach without allowing the cow too much room to walk onto the feeding surface. In addition, cows should have unrestricted comfortable access to clean and good quality drinking water. Providing rubber matting next to the feeding surface encourages feed intakes because cows will stand there for longer periods.

Points to observe include:

- Do cows have bumps on their necks? (Bumps can indicate that the neck rail is too low or cows are straining to reach the feed.)
- Do they have swollen knees? (This indicates cows are reaching for the feed.)
- Is the feeding surface pitted and black with rotting feed?

10.4.4 Flooring

When cows are not eating, resting or being milked, they are standing or walking around the shed. Good traction is essential so they can move without slipping and injuring themselves, such as when they display heat. In addition, the merits of floor comfort are important around the feeding area, in any holding yards and in the milking shed. Floors should be designed to provide sufficient traction without being too aggressive, as well as cushioning where cows have to stand for lengthy periods.

Floor surfaces should have a finish with sufficient texture to provide traction as the cow walks but smooth enough to clean properly. There is a fine balance between traction and cleanability. A good floor surface should be finished with a texture, such as grooves formed in concrete either during construction or after it has cured. Manure removal equipment should not polish the surface. Rubber matting is often advantageous in high traffic and turning areas.

The condition of cow's hooves is also important, before they are put in the shed. They should not be newly trimmed just before being moved to a shed with new concrete floors, as they will still be tender.

Behaviour to observe includes:

- Do cows slip and slide? This is an indication of slippery floors.
- Do cows show caudal licking (to groom the fold between the udder and the legs)? This is an indication that cows are quite comfortable with their footings.
- Are regular heats not observed? This could be an indication that the floor is so slippery that cows do not exhibit normal behaviour.
- Do cows walk tentatively? The floor may be too slippery or if made of wooden slats, the slats may not be stable or cows have become lame because the floors are too rough.

10.4.5 Lighting

Increased lighting can increase milk yields; cows exposed to 16 to 18 hr/d of light followed by 6 to 8 hr/d of dark can increase their milk yield by 8 to 10%. In additional, animal movement can be distracted by shadows, light and dark areas or obstructed vision. Additional distractions affecting animal movement include:

- sparkling reflections on puddles or reflections of smooth metal
- chains that jingle or metal clanging and banging

- high pitched noises such as air hissing
- air draughts blowing towards approaching animals or fan blade movement
- clothing hanging on a fence or piece of plastic blowing
- people moving up ahead or small objects (such as a coffee cup) or a drain grate on the floor
- sudden changes in colour of equipment, particularly colours with high contrasts
- race entrances that are too dark, but stock will move from a dark to a brighter place. However, they will not readily move towards a blinding light such as the direct sun
- stock will baulk at one-way and backup gates, which should be installed a few body lengths away from the crowd pen, and should be operated by remote control ropes.

In addition to the above five zones, there are other physical features around the shed that can affect cow behaviour, such as shed layout or effluent removal system. Dairy managers and staff should intuitively think about cow comfort in all these zones because seeing life 'from the perspective of the cow' is an essential attribute of a good dairy farmer. In other words a good dairy farmer develops the ability to 'think like a cow'.

Appendix 2 presents pictures of a dairy cow in different health and behavioural states, first a healthy and happy cow and second a sick and distressed cow. In addition to the pictures, Appendix 2 includes a list of observations and interpretations of various parts of the cow's anatomy.

10.5 Managing surplus stock to the dairy herd

Dairy farms can generate income additional to the sale of raw milk and value-added dairy products. These include:

- cull cows
- excess dairy stock, such as bull calves
- dairy stock specifically grown out for beef, called dairy beef
- manure
- excess fodder
- grass cuttings for planting material on other dairy farms
- biogas, from dairy effluent.

This section deals specifically with managing stock surplus to the milking herd.

10.5.1 Cull cows

Milking cows are normally sold when the farmer considers them uneconomic to try and mate to produce milk in future years. For example, they could be culled from the milking herd because of acute animal health problems, such as acute trauma or acute sickness. They could also be culled because of chronic disease problems, such as infertility, lameness or mastitis. They may be selected for sale because their milk production is minimal. They may be sold because the farmer needs the space or the feed for more productive stock. He may also need additional finances to purchase other stock or for personal reasons, such as a family celebration.

If they show symptoms of acute disease, care must be taken to reduce the likelihood of that disease spreading to other stock on the farm. They should be isolated from all stock and when transported to the saleyards, the vehicle should be thoroughly cleaned and sterilised. The major objective should be to remove them from the farm as rapidly as possible.

It is unlikely that farmers will want to improve the body condition of any cows culled because of chronic animal health problems, although the farmer may want to keep them on the farm if their sale price is likely to improve in the foreseeable future. If they are still milking, their diet should be that of other milking cows and for, say infertile, cows – they can milked for as long as the farmer considers their returns from milk are greater than the feed and other farm costs they may incur. If non-lactating, they should be fed sufficient nutrients to at least maintain their body condition, so their sale price is still maintained.

If cows are culled because of acute injuries, such as a broken limb or dislocated joint, they should be sold quickly and with as little additional stress as possible to ensure their animal welfare is not compromised. Farmers should, and mostly do, fulfil their obligations to the wellbeing of their livestock.

Another class of cull stock are young calves that have a history of chronic or acute health problems and are unlikely to ever become productive cows in the herd. Such decisions are value judgements, but there is much evidence that such heifer calves are likely to be culled from the milking herd much earlier in their life than would calves that had not suffered growth setbacks in their early life. It is worthwhile recording which calves suffer from chronic disease in early life and noting their longevity in the herd. This information will assist farmers with their value judgements as to which of these calves should be culled. Selling them just passes the problem onto other farmers, so it would be better to euthanase them rather than transport them to the sale yards.

10.5.2 Excess dairy stock

Low-producing cows would also be classified as excess dairy stock, but this class refers more specifically to male calves and potentially poor-performing heifer calves. These are normally sold within a week or so of birth.

In Australia, animal welfare regulations stipulate that these animals must be more than 4 days old and should weigh more than 23 kg before they can be transported. Newborn calf age is usually determined by the state of the umbilical cord, or 'the string' as farmers frequently call it. If it is moist, it is likely that he calf is less than 4 days old.

10.5.3 Dairy beef

Growing out bull calves for slaughter at 18 months or older, is frequently a major generator of income for many dairy farmers. These stock can be kept entire and grown out as bulls, or they can be castrated at an early age and grown out as steers. Bulls grow faster than steers but when they reach puberty, they can become a nuisance with milking cows on heat. Target growth rates should be at least 0.5 kg/day. There may be additional recommendations with regard to vaccinations and other health management issues, and such information should be sought from local animal health specialists.

Purebred or crossbred Friesians are preferred over Jerseys or local breeds and generally return more money per kg live weight when sold for dairy beef, so the breed type of the bull calf should be considered when deciding on its future fate on the farm. On the whole, milk generates more income per kg fresh grass grown on the farm than does beef, so it is important to 'do the sums' on dairy beef production when farm stocking rates are high. Acceptable growth rates are unlikely, if the major part of the ration is very low quality forages, such as rice straw. As with milking cows, green grass is the best forage to feed dairy beef animals.

It is important to plan the finishing (or fattening) strategy for dairy beef. Target live weights should be used in any feeding program, so using a chest girth tape to estimate live weight is worth considering. Friesians are later maturing than other dairy beef animals, meaning that a higher level of concentrate feeding will be required to produce a suitable sale animal, namely one with a good degree of finish (or some cutaneous fat cover over the ribs).

11

Introducing the concept of farm business management

This chapter presents a *modus operandi* for managing the day-to-day, medium and long-term operations of a successful and profitable dairy farm, small or large.

Dairy farmers are business managers, irrespective of the size of their milking herd. A successful business is based on a good understanding of the technology underlying the production of the end product, in this case raw milk, the ability to run the day-to-day operations at a profit and to make astute decisions regarding investments in its sustainable future. The scale of operation is generally limited by personal asset worth, but even smallholder farmers can make good financial returns on their dairy enterprise.

Smallholder dairy (SHD) farmers often do not have business support so have to learn, by trial and error, the best decisions to make relevant to their dairy enterprise. This chapter will provide them with a *modus operandi* on which to base their business decision-making processes. Large-scale dairy operations would generally have full-time administration staff with accounting skills to support the Farm Manager and senior management team.

Every day we all manage our personal assets when we make household business decisions at the shops, schools and in the wider community. It is the same with the dairy farmer when deciding on today's livestock feeding program, next week's crop agronomy program or the optimal herd size for next year's likely farm gate milk price. Such decisions are based on the elementary frameworks of farm business management.

Most farmers intuitively think about farm costs and returns. However, greater use should be made of ways to make them become aware of the relative importance of all their financial inputs, in terms of their contribution to the cost of production (COP) per kilogram or litre of milk produced on the farm. In addition, when contemplating changes in their routine farm practices, such proposed changes should be appropriately costed to allow farmers to make more meaningful and timely decisions.

The performance and sustainability of any dairy value chain in the tropics depends on the continued supply of raw milk. Economic pressures, such as those experienced by dairy industries throughout Asia, require each farmer to be more aware of their individual COP. Without such skills, farmers cannot prioritise their management decisions to address the high cost items of their production systems.

In addition, better knowledge of farm business management allows support organisations to more clearly define the key drivers of profit on dairy farms. This information can be used to develop regional and national strategies for government departments and national dairy organisations, such as those overseeing the producer-driven dairy cooperatives, to routinely evaluate and update their industry policies.

For the sake of simplicity, the following discussion uses the male pronoun (he) rather than both gender pronouns (he and/or she). It is clearly understood that on many farms, business decisions are jointly made by the husband and wife team.

11.1 What is good farm business management?

So what encompasses good farm business management for dairy farmers? Makeham and Malcolm (1986) summarised the essence of successful small and largeholder farming in the tropics. Of the eight key areas of knowledge they listed, five utilise skills in business management.

They list these key areas, with the three non-business management areas in italics, as follows:

1. *Crop production and protection.*
2. *Animal production.*
3. *Machinery selection and maintenance.*
4. Economic aspects of farm management (COP).
5. Credit and finance.
6. Marketing.
7. Managing labour and communications.
8. Information gathering.

From lists such as this, a set of key task areas for good dairy managers can be developed, and within each task area progress can be quantified through developing a series of Key Performance Indicators (KPI). The following example list of key tasks are for three major management areas on any tropical dairy farm:

Production technology

- Prepare land then plant, fertilise, weed, water (in some situations) and protect the crop, likely to be a forage crop (see Figure 11.1).
- Harvest, store and market the crop (through livestock rather than in the marketplace) to get the best return with minimum waste.
- Feed animals properly, prevent disease outbreaks and recognise disease symptoms (see Figure 11.2).
- Achieve high reproductive and survival rates.
- Obtain or produce nutritionally correct feed at the optimum (generally lowest) cost.
- Provide the right housing for effective production, protection, hygiene and harvesting of the animal product (see Figure 11.3).
- Where machinery is involved, be able to choose the most appropriate types for the job, ensure they are properly maintained and serviced and, when necessary, find a good mechanic.

Figure 11.1: Throughout the tropical world, Napier (Elephant) grass is the fresh forage of choice for feeding dairy cows in large or small herds. This is a picture of Napier grass grown along the roadside in Kenya, with minimal fertiliser input, either from manure or urea.

Figure 11.2: Biosecurity is essential in the tropics because of the presence of so many wandering stock and contagious diseases. This is a picture of the truck tyre water bath at the entrance to a North Vietnam dairy feedlot.

Figure 11.3: Free stall sheds do not have to be large to be effective. This Vietnamese version is built for a smallholder's herd of just 6 milking cows.

People skills

- Have harmonious relationships with farm workers by giving them a reasonable amount of responsibility.
- Be interested in the welfare of people working with the farmer.

- Establish a clear chain of command so each person knows to whom they are responsible and so does not have several bosses telling them what to do.
- Set up a system of supervision to ensure the work done is of a proper standard.
- Create a system of communication and involvement, so that all know what progress is being made in achieving goals and objectives of the farm operation.

Business management

- Use specialist advisers to help analyse the important production and financial aspects of the farm business.
- Through appropriate records, and other relevant information, be able to work with an adviser to produce annual farm plans, together with budgets, aimed at producing as much food and money as they need or have the ability to do.
- Prepare plans of action in case of abnormal seasons and/or price.
- Plan well in advance so that all inputs are available when required, and in correct quantities.
- Prepare physical and financial reports at regular intervals that are timely, accurate, relevant, brief and clear for the persons who control the farm.
- Determine the most favourable forms of credit that can be obtained for different activities.
- Develop good honest working relationships with bankers, financiers or other credit managers.
- Be able to prepare realistic applications and finance budgets to obtain such credit.
- Have the ability to know when borrowings are too great to be repaid from farm income.
- Assess the different ways of preparing and selling the farm products.
- Work out the best way(s) of marketing (assembling, preparing, transporting, selling) to return the greatest long-term benefit.
- Be able to obtain relevant information on any problem quickly. Information sources could be other successful farmers, extension agents, private agribusiness companies, research workers, libraries, the internet, teachers and friends.
- Develop effective thinking and reasoning skills that should be combined with common sense and even mini 'trial and error' experiments.

11.1.1 Recording business data in the farm office

The importance of maintaining good farm records cannot be overstressed. This is much easier in a farm office. An area at home or in the dairy shed should be dedicated to keeping records. It must have a desk and good lighting for night-time work. It must be a quiet place to set up the office files (preferably in a filing cabinet) and computer (if the manager has or needs one) and office supplies. He will need a system of storing and easily accessing all the financial paperwork. These include files relating to farm production (milk yields, veterinary reports, other stock and forage crop production data) and for each of the vendors (feed suppliers, veterinarians, cooperatives etc.), creditors, milk supply centre and any other farm-related agents. It is preferable to separate files from the dairy enterprise with those from other farm enterprises and it is important to separate these business files from any personal financial files. A file will need to be kept on unpaid bills. A simple recording system for payment of bills (with details of how it was paid) and confirmation receipts from sale of farm produce should be developed.

The 'how and when' of keeping farm records depends on the person recording them. Computers are very convenient but require money to purchase and skills to operate efficiently. As computers can break down, 'hard' (paper) backup copies should be routinely made. Record keeping should be given as high a priority as other farming activities, so it should not be 'put off' until the last job each day when simple bookkeeping mistakes can more easily be made. This topic has been further discussed in Chapter 7 at Section 7.1.

11.2 Breaking down costs and returns on dairy farms

The actual costs of producing milk on SHD and large-scale farms can be broken down into two major components:

1. Variable (or direct) costs that are directly related to the farm's milk output and so to the amount of variable inputs, such as fertilisers, purchased concentrate and forages, and herd costs.
2. Overhead (or fixed) costs that are not directly related to the amount of milk sold by the farmer, as they must be paid whether anything is produced. These include land rent, government land taxes, loan repayments and other finance costs, and living expenses. Labour costs can be categorised either as overhead or variable, but in this chapter all will be considered as overhead costs.

For the purpose of calculating the COP of milk, costs can be broken down into four categories, two variable and two overhead (Moran 2009a) as follows:

1. **Variable** costs. These are broken down to:
- herd and shed costs, to maintain the entire dairy herd and to harvest the milk
- feed costs, to feed the milking herd.

The more milk produced and the bigger the dairy herd, the greater these variable costs.

2. **Overhead** costs. These are broken down to:
- cash overhead costs which involve actual payments, such as for employed labour and interest on borrowed money, rates and other farm administration costs

- imputed overhead costs, or hidden costs because no cash changes hands. Family labour is the classic example where the farmer and his family work the farm but all too frequently don't pay themselves for their labour. Depreciation of farm equipment is another imputed cost that becomes obvious when the equipment must be replaced.

The question often arises, what is the farm manager worth to the business? The answer is either what he could earn if he spent that time being paid to do other work (that is the opportunity cost of his farm labour), or what it would cost to employ someone else to do his job. With regard to the latter, as it requires more skills to manage a large dairy herd (say 500 cows) than a small one (say 20 cows), the bigger the herd and the more complex the job, the greater should be the manager's/operator's allowance or imputed labour costs. This is rarely considered in Asian smallholder dairy farming.

This COP analysis includes finance costs on borrowed money. It is the choice of the farm business adviser as to whether interest on loans is included as a cost of producing milk on Asian smallholder farms. Because they can constitute a major cash outlay each year for smallholder farmers, they should be included in his annual financial commitments and future farm budget projections. Consequently, they have been incorporated into the following COP analyses. This decision highlights the importance of clearly describing the particular components of any financial analyses so the reader is clear as to exactly what is and what is not included in the final 'bottom line' measure of COP. Unfortunately, this is rarely the case, leaving the reader unsure as to how such data can be interpreted and importantly, compared with other COP estimates undertaken by other dairy farm management specialists, probably using different methodologies.

Fuel and oil are normally included in the feed costs although some farm economists consider repairs and maintenance of farm machinery as an overhead rather than a variable cost. It doesn't greatly matter as long as it is only included once.

As dairying is frequently just one of the enterprises on many smallholder farms, it is important to only consider the costs relevant to and the income generated from the dairy enterprise. Such apportioning of farm finances is often not easy because labour units, machinery and farm facilities are frequently used for a diversity of farm enterprises. In addition, if feed for any dairy animals (young stock as well as adult cows) is produced from a cropping enterprise on-farm, such as rice straw or maize stover, it should be given a cost to the dairy enterprise.

The total COP is then the sum of all farm costs included in Table 11.1 and Figure 11.4. Unfortunately one still finds published estimates of COP for SHD operations that do not include family labour and finance costs. These create a false assessment of the true costs of dairy farming and if used to base government

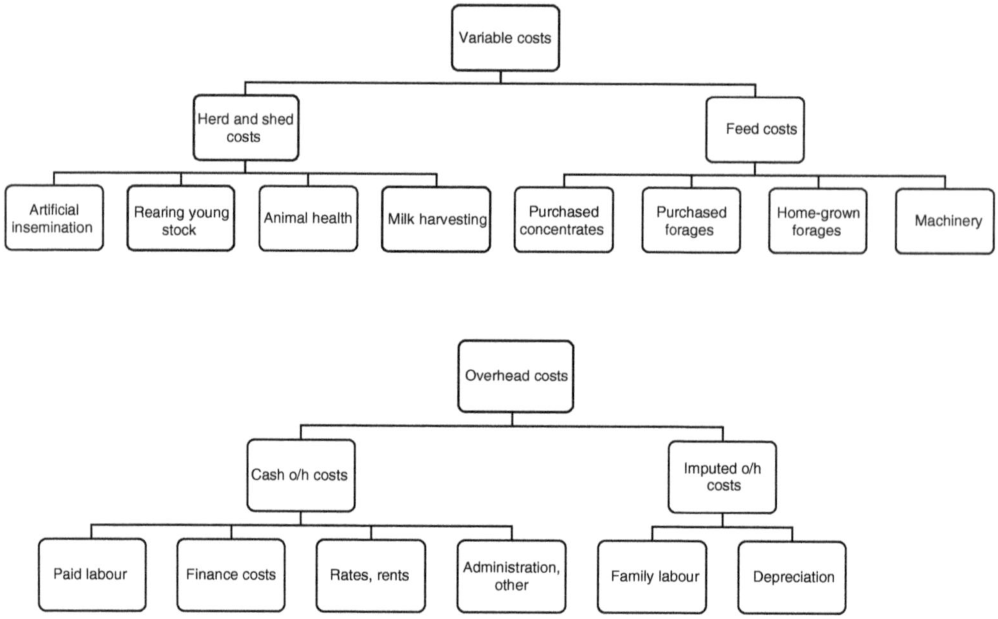

Figure 11.4: Farm costs for smallholder dairy farms.

policies for dairy development and even milk prices (as in some countries), they do not paint the true picture of the economics of smallholder dairy farming.

11.2.1 Quantifying farm profit

In order to improve profit, it must first be measured. Most farmers think of profit in terms of cash or money left over from income after deducting all the costs involved in earning that income. In other words, to them profit generally refers to some surplus of income over costs, or in economic terms, the difference between the gross income and the operating costs. This may be the simplest measure of profit, but it is not necessarily the best. Profit can be expressed in three ways.

1. Cash. Does the farm generate enough cash to pay the bills, repay the loans and reward the farmer for his work? This can be expressed by a range of indicators such as cash operating surplus, milk income less feed costs, milk gross margin or economic farm surplus.
2. Efficiency. How efficiently are the farm resources being used? For a general overview of the business performance, this is expressed as return on assets, however, for a more detailed assessment of what the farmer actually owns (namely his equity), a more suitable measure is return on equity.
3. Wealth creation. Does the farmer own more than he did last year? This is expressed as capital gains or more suitably as the difference between the two years' equity.

Table 11.1. Categorising farm costs and income on smallholder dairy farms.

Category		Details
Variable farm costs	Herd and shed costs	• Artificial insemination; inseminator, semen, drugs associated with reproductive management • Young stock; raw milk or calf milk replacer, concentrates and roughages and herd management to point of calving • Animal health; veterinarian visits, drugs, vaccines and drenches • Milk harvesting; rubber liners, detergents and sanitisers, maintenance of milking machines, hot water, transport to milk collection centre, cooperative commission
	Feed costs (for milking and dry cows)	• Purchased concentrates; formulated or ingredients • Purchased forages; grass, roughage by-products • Home-grown forages; fertilisers, irrigation, processing/storage, weed and pest control • Machinery; fuel and oil, repairs and maintenance
Overhead farm costs	Cash overhead costs	• Paid labour • Farm rates • Farm administration and insurance • Finance costs; interest, bank fees • Other; such as telephone, professional advice, office equipment, postage
	Imputed overhead costs	• Family labour, such as operator's allowance • Depreciation
Farm income	Milk sales	• To milk collection centre or milk processor (formal marketing) • Direct to consumer or through milk vendor (informal marketing) • Consider value-adding milk
	Non-milk sales	• Sale of excess stock (bull calves) and salvage value from culled milking cows • Sale of forages • Sale of manure as fertiliser or biogas as energy source

Quantifying cash and non-cash profit

The simplest measure of cash profit is cash operating surplus (COS), which quantifies the sum of all the cash flows on the farm.

$$\text{COS} = (\text{farm cash income}) - (\text{farm cash costs})$$

Milk income less feed costs (MIFC) is a useful measure of cash profit because it is relatively easy to measure and provides a guide to how well the cows are being fed. It does not take into account the costs of feeding the non-productive stock on the farm, namely the dry cows and replacement heifers. MIFC is also called Feeding Profit.

$$\text{MIFC} = (\text{milk income}) - (\text{feed costs for milking cows})$$

Another way to quantify cash profit on dairy farms is using the milk gross margin (MGM). This calculates the income from milk sales less the variable costs to produce that milk.

$$\text{MGM} = (\text{milk income}) - (\text{variable costs})$$

The most sophisticated methods to quantify farm profit, first uses non-cash farm income (changes in stock and land values) to calculate gross farm income (GFI) and then non-cash farm costs (imputed labour and depreciation) to calculate net farm income (NFI).

$$\text{GFI} = (\text{Total farm cash income}) + (\text{changes in stock inventory})$$

$$\text{NFI} = \text{GFI} - (\text{variable + overhead cash costs \{excluding finance costs\}}) + (\text{imputed costs})$$

Net farm income is also known as economic farm surplus (EFS) or operating profit. Operating profit does not include finance costs because these are the cost of acquiring the services of the assets used, and are not directly related to their performance. Earnings Before Interest and Tax (EBIT) is sometimes used as a measure of NFI, this being a relatively new term in farm management economics, and is defined as farm revenue less farm expenses before the payment of interest on loans and income tax.

There are three different descriptions of profit that require some explanation. *Operating profit* describes the return to all capital and, by removing finance costs, it becomes the return to the farmer's equity, so quantifies his *net profit*. Finally, subtracting his income tax from the net profit quantifies his *growth in equity* or addition to wealth.

Break-even milk price is another way of expressing farm profit. It is the indicative milk price required to cover the cash costs of production, but not principal repayments and any capital expenditure. It also excludes other farm income, just dealing with milk production. When compared to the farm gate milk price received, the difference provides a measure of the profit or loss incurred by the farmer on a per kg milk basis.

Quantifying change in farm efficiency

The term capital refers to all the production resources of the farmer. The most important are land, buildings, improvements (such as built-up soil fertility and irrigation), machinery, stock, fuel, labour management skills and credit. Many of these could be converted to cash by selling them. The cash sum available from their sale is the farm assets. After paying off any debts owed on the farm, they are the farmer's own capital, his net worth or his equity. Total assets are a good measure to compare the business performance of that farm with others of similar size of operation, however, a more meaningful measure for that farmer relates to his total equity in the farm.

The market value of the total resources on the farm is sometimes known as the total capital of the farm. This is calculated by summing the market value of the land, improvements and animals, plus the machinery and feed reserves. With the numerator net farm income, return on assets (ROA) is calculated as follows:

$$\text{ROA (\%)} = \frac{\text{Net farm income}}{\text{Assets}} \times 100$$

This calculation takes no account of debts owed so quantifies the earning rate of the total bundle of resources employed in the business. In practice, the farmer has to manipulate the total resources under his control, not just those that are debt free. ROA provides a guide to those responsible for the use of capital (this could be an individual, a cooperative or a government department). It also allows the performance of this capital, invested as it is, to be compared with alternative possible investments.

As the farmer probably does not own all the farm assets, he is more interested in how efficiently he is using his own assets. The farmer's equity is calculated by adding the market value of all the resources he owns then subtracting it from a total of all the money he owes (his liabilities). Equity, expressed in monetary terms quantifies the net worth of the farmer. However, it is usually expressed as a percentage, calculated as follows:

$$\text{Equity (\%)} = \frac{\text{Assets} - \text{Liabilities}}{\text{Assets}} \times 100$$

Expressing the farm's annual profit, after paying interest and taxes, as a percentage of this capital is one measure of the effectiveness of the management of the farm's resources. As the calculation does not take into account any debts, it must also exclude the finance costs associated with these debts. The return on equity (ROE) is then calculated as follows:

$$\text{ROE (\%)} = \frac{\text{Net farm income} - \text{finance costs}}{\text{Resources owned}} \times 100$$

The ROE measures the farmer's effectiveness as a combination of annual inputs such as labour, irrigation, fertilisers, machinery and other resources used to operate his dairy enterprise. It also quantifies the rate of earning of his capital committed to his farm relative to the rate of earning if it were used in some other income-generating enterprise.

Calculating his ROE shows the farmer how efficiently he is running the annual operations of his farm business. If it is very low, the farmer should consider alternatives by asking the questions:

- Can he increase his ROE by using better farming methods, borrowing extra money to improve production or diversifying his farm enterprises?
- Should he transfer his capital from this farm and move to a different locality where the ROE is likely to be higher?

- Should he sell up and move into another form of investment?
- Is his ROE low because there has been a large increase in the value of his assets?
- Should he use his increased equity to borrow more money to further develop his farm and earn more income?

The more rapid the annual increase in asset worth, the more difficult it is to maintain a constant ROE. An increase in asset value provides more collateral against which to borrow to invest in the operation to increase farm income. It can also mean an increase in land tax, hence greater farm costs.

Farm profit can be quantified in various ways (see Chapter 11 of Moran 2009a). These are presented as a flow chart in Figure 11.5.

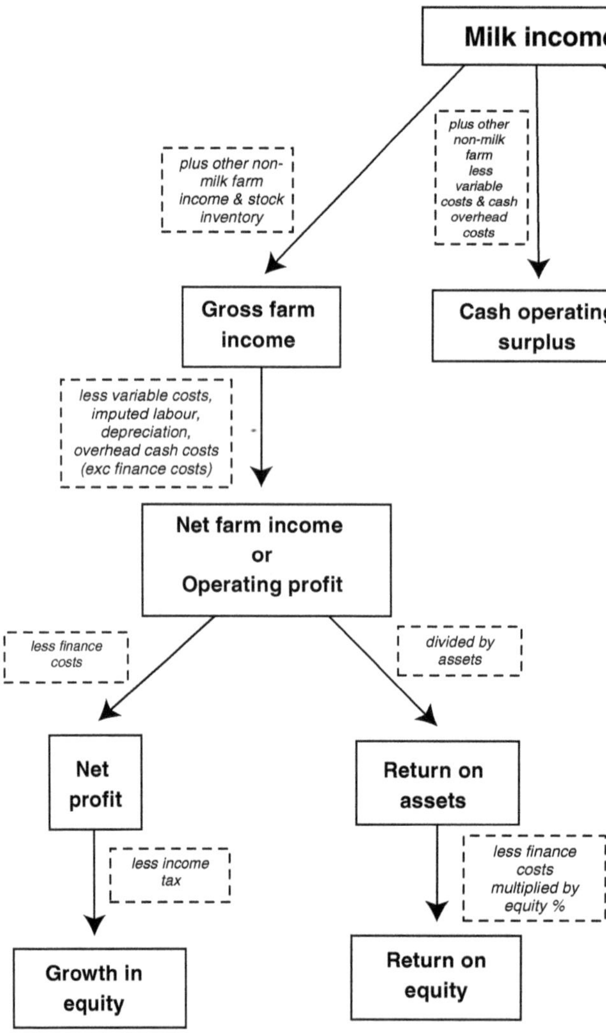

Figure 11.5: Flow chart of measures of farm profit on dairy farms.

11.2.2 Questionnaire for farm business analyses

Appendix 3 contains a questionnaire for the collection of farm data to calculate a full COP calculation and subsequent profit analysis on any farm, large or small. In addition to the necessary data on costs and returns on the farm, the questionnaire includes other useful questions on specific aspects of the farm management.

11.3 A case study of the herd dynamics and farm profits achievable on smallholder dairy farms in Malaysia

11.3.1 Impact of cow milk yields on farm profits

Farm production and business performance data were collected from 30 dairy farms in Peninsula Malaysia during September 2012 by Moran and Brouwer (2013a). Observations of the stock, cowshed, farm facilities and forage production area were made to assess current farm practices and the general state of the stock and the supporting dairy infrastructure. Farmers were interviewed about key aspects of their farm management, the costs of farm inputs and their herd performance to develop a series of KPIs. The business focus covered specific aspects of milk returns and feeding management to calculate total feed costs, feed efficiencies and farm profits. Gross farm profits were calculated, including and excluding imputed labour costs. The farms were split into three groups to quantify the impacts of farm management on cow milk yields and farm profits. The key data findings are summarised in Table 11.2. The unit of currency in the table and following discussion is the Malaysian ringgits or MR, which in September 2015 had an exchange rate of 4.25 MR/US$ or 3.05 MR/A$.

Herds with higher average milk yields contained a greater proportion of adult cows and replacement heifers. The milking cows had higher feed intakes and higher ration quality while the cows had higher feed efficiencies in that they converted more of their feed into saleable milk. Even though the farmers spent more money on feeding their milking cows better, these more productive herds yielded greater feeding profits from milk sales and they had lower costs of unit milk production. The efficiency with which these farmers utilised the farm assets (both gross assets and assets that they actually owned) increased as the cows produced more milk.

The survey provided many valuable insights into why some farms are productive and profitable and why others are not. In essence, higher per cow milk yields and farm profitabilities were recorded on farms that were better equipped and better managed. The more productive and profitable farmers had more reliable electricity and water supplies, provided specific calving down areas, did not graze their milking cows and did not suckle their calves on milkers. In addition, they used artificial insemination rather than natural mating, used calf milk replacer as part of their milk-rearing program, routinely used dry cow therapy as part of the mastitis control program, kept farm records and had fewer problems with mastitis, lameness and young stock rearing. More of the cows on the most profitable farms had high peak milk yields and fewer had short lactations. Although they invested more in feeding for their milking cows, the resultant greater feed conversion efficiencies on these farms yielded higher feeding profits and higher returns on total farm assets and equities.

Table 11.2. The impact of herd average daily milk yield on farm performance and business data of 30 farms in Peninsular Malaysia. The farms are grouped into either A, B or C (10 farms per group) based on increasing per cow milk yields.

Farm data	A	B	C	Sig
Herd average daily milk yield (kg/cow/d)	7.5	9.7	12.4	*
Size of milking herd (cows)	22	48	27	
% milking cows in adult herd	49	53	61	
% replacement heifers	47	73	80	*
Dry matter intake (kg/milking cow/d)	10.8	12.4	14.6	*
Ration metabolisable energy content (MJ//kg DM)	8.1	8.5	9.0	*
Ration crude protein content (%)	11.6	12.1	12.3	
Feed conversion efficiency (kg DM/kg milk)	0.70	0.82	0.87	*
Total feed costs for milkers (MR/cow/d)	7.44	8.75	11.41	*
Total feed costs as % milk income	78	76	49	*
Milk income less feed costs for entire herd (MR/kg milk)	0.53	0.57	1.24	*
Gross farm profit (MR/kg milk)	−2.01	−0.75	−0.05	*
Cost of production (MR/kg milk)	4.77	3.53	2.82	*
Feed costs (% total farm costs)	38	40	43	
Return on assets (%)	−0.6	−0.4	0.1	*
Return on equity (%)	−0.9	−0.5	0.1	*

MR, Malaysian ringgits.
* Significant difference between herds.

Cowshed designs were generally poor in that roofs were low, shed hygiene had much that could be improved and fans and cooling sprinkler systems were virtually non-existent on any of the 30 farms. In addition, many of the farms suffered from a lack of productive cows in their herds. Future herd management must concentrate on improving reproductive performance and in some instances, reducing young stock mortality as well as improving the nutritional status, hence performance, of the milking herds. Of the 30 farms surveyed, only eight had positive gross farm profits, although this increased to 18 farms if farmers excluded their family labour from the costs of milk production.

11.3.2 Financial benefits of improving herd and feeding management on Malaysian dairy farms

One of the best ways to encourage farmers to consider ways to improve their current management is to demonstrate the economic benefits from such practice changes. Below are five examples (Moran and Brouwer 2013b) of such savings, calculated from economic data derived from this Malaysian survey.

1. Reducing age at first calving by 6 months (from 33 to 27 months)
   Additional income: For a milking cow, milk income less feed costs is RM 1.45/kg milk or RM 14.50/d for a milking heifer producing say 10 kg/d or RM 2610 over 180 days.
   Additional costs: Daily feed costs for a milker is RM 9.20/d while for a yearling, it is RM 2.46/d, or RM 6.74/d difference. Over 180 days, this amounts to RM 1213.
   Net profit: RM 2610 additional income less RM 1213 additional costs amounts to **RM 1397**/heifer profit when calving 6 months earlier.

2. Reducing calving interval by 3 months (from 17 to 14 months)
   Additional income: For a milking cow, milk income less feed costs is RM 1.45/kg milk or RM 17.40/d for a milking cow producing 12 kg/d or RM 1566 over 90 days.
   Additional costs: Daily feed costs for a milker is RM 9.20/d while for a dry cow, it is RM 4.54/d, or RM 4.66/d difference. Over 90 days, this amounts to RM 419.
   Net profit: RM 1566 additional income less RM 419 additional costs amounts to **RM 1147**/cow profit when calving 3 months earlier.

3. Increasing the lactation length by 2 months (from 8 to 10 months)
   Additional income: For a milking cow, milk income less feed costs is RM 1.45/kg milk or RM 17.40/d for a milking cow producing 12 kg/d or RM 1044 over 60 days.
   Additional costs: Daily feed costs for a milker is RM 9.20/d while for a dry cow, it is RM 4.54/d, or RM 4.66/d difference. Over 60 days, this amounts to RM 280.
   Net profit: RM 1044 additional income less RM 280 additional costs amounts to **RM 764**/cow profit when milking for an additional 2 months.

4. Reducing the costs of calf rearing by replacing raw milk with calf milk replacer (CMR)
   CMR powder costs RM 6720/t and makes 7690 L CMR solution (at 130 g/L solution), therefore, costs RM 0.87/L solution. Raw milk sells for RM 2.40/kg or RM 1.63/kg more. For a milk-fed calf fed 2kg/d of milk for 12 weeks (or 84 days), using CMR provides **RM 274**/calf savings.

5. Reducing the costs of calf rearing by replacing calves suckling cows for 9 months cost with 4 months of feeding CMR
   From 4 above, CMR costs RM 0.87/L solution compared to RM 2.40/kg for raw milk. Assuming a suckling calf will drink 3 kg raw milk/d for 9 months, or 810 kg over 9 months, this costs RM 1944/suckling calf. By rearing the same calf on 2 L/calf/d of CMR solution and weaning it at 4 months, total CMR costs are RM 146/calf. The daily feed costs of yearlings are RM 2.46/d or RM 369/weaned heifer over 5 months (although this would be for yearling heifers). Total feed costs are then RM 515/CMR reared calf versus RM 1944 for the suckling calf, or **RM 1429**/calf savings. There may be a difference in live weight at 9 months of age between these two calves because of their vastly different rearing programs.

The economic data generated in farm surveys such as this one in Malaysia provides ideal opportunities to more objectively assess how improvements in farm management can be better quantified in terms of improved farm profits.

11.4 A case study of the Feeding Profits achievable on a large-scale intensive dairy farm in Indonesia

Intensive feedlotting of dairy cows is an expensive exercise, but when well managed, it can be very profitable. Once the facilities have been constructed, the stock acquired, the feeding systems developed (that is establishing the forage production area and the logistic of sourcing the ration ingredients) and the day-to-day feeding and herd management set in place, well-fed dairy cows will definitely return a profit. However, to develop a production system up to this point takes several years and a big financial investment. Thereafter cash flows will become positive so the initial investment will be recouped. Once the system has reached some degree of stability, with regard to cow numbers and milk throughput, hence positive cash flows, that is the most appropriate time to assess its long-term profitability.

From the previous section, profitability was quantified in various ways, such as return on investment, net farm income, cash operating surplus, milk gross margin or milk income less feed costs (or feeding profit). The following case study uses Feeding Profit as the measure of profitability. It is based on a large-scale intensive dairy feedlot in Indonesia. For this exercise, the Indonesian unit of currency is the rupiah (Rp) which had an exchange rate in Sep 2015 of 14 300 Rp/US$ or 10 373 Rp/A$.

11.4.1 Describing the farm and the stock

Prior to assigning costs and returns in this case study, a series of assumptions have been made describing the large-scale intensive dairy farm. These were:

- The farm was an established one with a stable number of livestock and an established source of regular forage supplies.
- It is located in the tropical highlands of East Java in Indonesia where the temperatures and humidities are not too stressful for the milking herd.
- The milking herd consists of 400 purebred Friesian cows made up of 100 cows in their first, 100 cows in their second, 100 cows in their third and 100 cows in their fourth lactations.
- Each 410-day entire lactation was split into four phases, namely 110 days in early, 110 in mid, 110 days in late lactation and 80 days dry.
    - During their 110 days of early lactation, cows were non-pregnant, losing 0.2 kg/day live weight and produced 125% of their average daily full lactation milk yield.
    - During the 110 days of mid lactation, cows averaged 1 month pregnant, maintained their live weight and produced 104% of their average daily full lactation milk yield.
    - During their 100 days of late lactation, cows averaged 5 months pregnant, gained 0.2 kg/day live weight and produced 78% of their average daily full lactation milk yield.
    - During the 80 days dry period, cows averaged 8 months pregnant.
- For cows at different ages:
    - First lactation cows weighed on average 500 kg and produced 4000 L milk over the 330 days, or 15.6, 13.0 and 9.7 L/day respectively in early, mid and late lactation, with a rolling herd average of 12.8 L/day.
    - Second lactation cows weighed on average 550 kg and produced 5000 L milk over the 330 days, or 19.5, 16.2 and 12.2 L/day respectively in early, mid and late lactation, with a rolling herd average of 16.0 L/day.
    - Third lactation cows weighed on average 600 kg and produced 6000 L milk over the 330 days or 24.4, 19.4 and 14.6 L/day respectively in early, mid and late lactation, with a rolling herd average of 19.2 L/day.
    - Fourth lactation cows weighed on average 600 kg and produced 7000 L milk over 330 days or 27.4, 22.8 and 17.1 L/day respectively with a rolling herd average of 22.4 L/day.
- For calculations on feed energy requirements, the milk contained 3.6% fat and 3.0% protein. Its initial unit sale price was 5500 Rp/L. Following value-adding, a second assessment of Feeding Profit was made with milk valued at 6500 Rp/L.

Actual farm data were collected on the local Indonesian costs and the nutritive values of the most common feeds likely to be fed in an intensively managed feedlot dairy in East Java. These have been called the 'preferred feeds' because they were assumed to initially form the basis of the feeding program at the dairy feedlot. Table 11.3 presents the assumed costs and nutritive values of these preferred feeds.

Table 11.3. Unit costs and nutritive value of preferred feeds for Indonesian intensive dairy system.

Feed category	Preferred feed	Unit cost (Rp/kg fresh)	DM (%)	ME (MJ/kg DM)	Crude protein (%)	NDF (%)
Forage	Young Napier grass*	250	16	8	12	60
	Maize silage*	600	28	9	8	50
	Rice straw	250	90	5	6	80
Energy rich	Formulated concentrate*	4000	90	12	16	25
	Maize grain*	3100	90	13	10	10
	Molasses	2700	79	13	4	1
Protein rich	Soybean meal*	6000	90	13	45	15
	Wheat pollard*	1700	90	12	15	30
	Brewers grain	800	22	10	25	55
Additives	Limestone*	500	100	–	–	–
	Salt*	1500	100	–	–	–
	Vitamin/mineral mix*	8000	100	–	–	–

* Feeds used in the following farm financial analyses.
Rp, Indonesian rupiah; DM, dry matter; ME, metabolisable energy; MJ, megajoules; NDF, neutral detergent fibre.

11.4.2 The Feeding Profit for milking cows

Using international nutrient requirements and feeding tables for milking cows (previously discussed by Moran 2005), various scenarios were developed of intensive feeding systems for dairy cows of different ages, live weights (LWT) and lactation milk yields. This allowed the calculations of how much of these 'preferred feeds' were required to form nutritionally balanced rations to achieve target daily milk yields at any particular stage of lactation. The calculations also took into account changes in the cows' body reserves (hence their LWT) and the growth of the foetus during pregnancy. The conventional way to quantify lactational changes in daily nutrient requirements of milking cows is to divide the lactation cycle into the four phases described above. The calculations were then based on cows with four ages or LWT and producing different volumes of milk in each of the three milk-producing stages of lactation.

The nutrient requirements and quantities of feeds required per cow per day were calculated from a computer program developed by the senior author called INDOFEEDPROFIT. This program is freely available to interested readers.

Computer programs for ration formulations (INDOFEEDPROFIT and INDOFARMPROFIT)

Several years ago, the senior author developed a computer program to calculate Feeding Profit. INDOFEEDPROFIT is a simple Excel spreadsheet in which the operator:

- Specifies the type of cow; namely its live weight, stage of lactation, pregnancy status, milk yield, milk composition and any changes in live weight over time.
- Develops a database of up to 17 feeds, with for each feed, its unit price in Indonesian rupiah (or any other selected foreign currency unit) and contents of dry matter (DM), metabolisable energy (ME), crude protein (CP) and neutral detergent fibre (NDF).
- Formulates a ration describing the number of kilograms fresh weight of each selected feed to provide sufficient nutrients to satisfy the cow's demand to achieve that predetermined level of cow production.
- Given the unit milk return (in this case 5500 Rp/L then 6500 Rp/L), the program then calculates the total feed costs, the total milk returns, hence derives the Feeding Profit for that scenario, in local currency units.

A second Excel spreadsheet, called INDOFARMPROFIT, was developed to summarise the INDOFEEDPROFIT various scenarios, based on the Table 11.3 'preferred feeds' ingredients, for both the milking cows and the growing heifers.

The programs allowed the selection of certain weights of fresh feeds and their formulation to provide sufficient ME then calculated the protein and NDF contents for each ration. The target protein contents were 16–18% in early, 14–16% in mid, 12–14% in late lactation and 10–12% during the dry period. The target NDF contents are 40% throughout lactation. Maximum DM intakes are based on 3% LWT, equivalent to 15 kg/day in first, 16.5 kg/day in the second and 18.0 kg/day in the third and fourth lactations. INDOFEEDPROFIT calculated the daily cost of each ration, then the associated Feeding Profit.

To ensure adequate roughage supplies in the ration, the amounts of maize silage and Napier grass were maintained at 10 and 20 kg/cow/day respectively during lactation and 10 and 30 kg/cow/day respectively during the dry periods. The allocations of energy sources, namely formulated concentrate and maize grain, varied from 2 to 6 kg/day and from 1 to 2.5 kg/cow/day respectively, with more being fed to the older, higher yielding cows. The allocation of protein sources, namely soybean meal and wheat pollard, varied from 1.5 to 2.5 kg/cow/day and 1 kg/cow/day respectively, to maintain protein levels at 15 to 16% throughout lactation to guard against possible protein deficiencies adversely affecting cow performance. The total feed costs also took into account the inclusion of additives in the ration, at the rate of 1.8% of the DM and costing on average for all the additives, 3360 Rp/kg. The resultant data are summarised in Table 11.4.

Table 11.4. Changes in average daily intake, feed costs, milk incomes and feeding profits of cows at different ages and stages of lactation in an Indonesian intensive dairy system. Calculations for entire lactation take into account the 80-day dry period.

Lactation	**Stage**	**DM Intake**	**Feed cost**	**Milk income**	**Feeding profit 1**	**FCR**	**Feeding profit 2**
		kg/day	000 Rp/day	000 Rp/day	000 Rp/day*	kg milk/kg DM intake	000 Rp/day*
First	Early	12.3	40.1	85.8	45.7	1.25	61.1
	Mid	11.4	35.5	71.5	36.0	1.12	48.9
	Late	10.9	33.1	53.3	19.9	0.87	29.5
	Entire	**11.1**	**31.9**	**56.5**	**24.6**	**0.94**	**34.7**
Second	Early	14.8	46.3	107.2	61.0	1.29	80.3
	Mid	13.9	42.2	89.1	46.9	1.14	62.9
	Late	13.0	38.8	67.1	28.3	0.92	40.4
	Entire	**13.1**	**36.7**	**70.7**	**33.9**	**1.00**	**46.7**
Third	Early	16.6	54.4	128.7	74.3	1.38	97.5
	Mid	15.7	50.3	106.7	56.3	1.21	75.6
	Late	14.3	44.7	80.3	35.6	1.00	50.0
	Entire	**14.5**	**42.9**	**84.7**	**41.8**	**1.08**	**57.1**
Fourth	Early	17.9	61.5	150.7	89.2	1.50	116.3
	Mid	17.0	56.5	125.5	68.9	1.32	91.4
	Late	15.2	47.9	94.0	46.2	1.11	63.1
	Entire	**15.5**	**47.2**	**99.3**	**52.1**	**1.19**	**69.9**

* Feeding profit 1 calculated for 5500 Rp/L milk return, Feeding profit 2 for 6500 Rp/L milk return.
DM, dry matter; FCR, feed conversion ratio.

Table 11.4 presents average daily data for the various ages and levels of milk production. It also presents average daily data for entire lactations, taking into account the dry period. In addition, it presents a measure of feeding efficiency called Feed Conversion Ratio or FCR, which is the amount of milk produced from each kg of feed DM consumed. It is also assumed that heifers first calved down at 28 months of age with calving intervals averaging 13.7 months, and were culled after four lactations at 6.9 years of age.

Clearly there is considerable money to be made from intensively feeding and managing cows in Indonesia, following the establishment period of such a new intensive dairy venture. Granted the Feeding Profit does not take into account all production costs but dairy specialists usually consider that feed accounts for 70 to

80% of the variable costs of such a system. At 5500 Rp/L milk return, the Feeding Profit increases from 25 000 to 52 000 Rp/cow/day as cows progressed from their first to their fourth lactations, while at 6500 Rp/L milk return, this increases from 35 000 to 70 000 Rp/cow/day. Once the system has stabilised, each milking cow then returns an additional 9 to 25 million Rupiah per lactation, depending on her unit milk return and lactation number. This is equivalent to Feeding Profits of 650 to 1800 US dollars per cow per lactation respectively.

These data highlight another very important point, **namely you get back in milk what you feed out in the trough.** Under such feedlot systems, cows fed only 11 to 12 kg/day of DM will only produce 4000 L over a full lactation, which means their highest milk yield is likely to be only 15 L/day with a rolling herd average of only 12 or 13 L/day. To achieve 5000 L over a full lactation requires feeding 12 to 15 kg DM/day while 6000 L requires 14 to 16 kg DM/day and 7000 L requires ensuring milking cows can access 15 to 18 kg DM each and every day. All too often, managers of conventional 'high input' dairy cow systems in the tropics can only achieve milk yields of 12 to 14 L milk/day in their cows purely and simply because they do not provide enough dry matter for their forever hungry milking cows. There are other factors in this, such as poor quality fodder (thus reducing potential appetite due to slow rates of feed passage through the stomachs), poor young stock management (thus producing stunted heifers) and lack of attention to cow comfort (namely providing a comfortable soft bed for cows to relax on and ensuring adequate shed ventilation).

11.4.3 Feed costs for rearing heifer replacements

For simplicity, young stock in this case study were fed on a mixture of Napier grass, maize grain and soybean meal with a target growth rate of 0.57 kg/head/day throughout their growing phase. Increasing amounts of Napier grass were fed as they grew (from 8 to 20 kg/head/day), as with maize grain (from 1.6 to 5 kg/head/day) and cottonseed meal (from 0.2 to 0.6 kg/head day). The protein content was maintained at 16% until they reached 6 months of age, after which it was reduced to 12%. They are being managed to conceive at 15 to 18 months of age.

Table 11.5 presents the feed intakes and costs to achieve 0.6 kg/head day growth rates from replacement heifers. This clearly shows that for an investment of only 10 000 to 23 000 Rp/head/day, dairy replacement heifers can be fed and managed to become productive highly productive milking cows. Furthermore, it is important to have heifers calving down at 24 to 27 months of age, rather than the 30 to 33 months of age that is all too common in tropical dairy systems, even the supposedly intensive ones.

Table 11.5. Live weights, feed intakes and feeding costs of dairy heifers growing at 0.57 kg/day.

Stage	Age (months)	Live weight (kg)	DM intake (kg/day)	Feed cost (000 Rp/day)
Weaned	3–6	141	3.6	10.6
	6–9	194	4.1	10.4
	9–12	246	4.9	12.1
	12–15	298	5.9	13.9
Pregnant	15–18	349	6.6	16.1
	18–21	402	6.9	16.7
	21–24	450	7.4	19.2
	24–28	502	8.5	23.5

The major conclusion that can be reached from the above case study is that well-managed, intensive dairy systems can be very profitable in humid tropical countries such as Indonesia.

12

Regional smallholder dairy development programs

This chapter presents an overview outlining the successes and failures of past regional development dairy programs for smallholder dairy farming in Asia. A five-step model program is then presented, highlighting all the elements of a successful farmer extension program.

With increasing demands for liquid and processed dairy products, virtually every country in tropical SE Asia has proactive programs to encourage and stimulate domestic production of raw milk on smallholder dairy (SHD) farms. To ensure success, such programs must be carefully planned and implemented. Unfortunately, this has not always been the case.

The public sector is very active in SHD development, and this can often discourage private sector investment. Furthermore, government direct investments in milk production and milk processing are often unprofitable. Public sectors should concentrate their efforts in dairy development by facilitating the business operating environment through policies and strategies via tailored institutions and technical support and also by promoting dairy trade and liquid milk consumption. This should encourage and facilitate private delivery of public services.

To produce milk competitively, smallholders need accessible and affordable complete packages of technical support services, such as animal health and AI services. The private sector must be fully engaged with both government and farmers in developing regional strategies and also national action plans. Regional dairy development is best undertaken by what is known as a Public Private Partnership or PPP.

The last 20 years of dairy research, development and extension in many Western countries has produced quite sophisticated dairy production systems. Herd sizes have grown, efficient feeding systems have evolved and many farmers routinely monitor test results on their cows for milk production, composition and quality and for mastitis. They then use this information for making decisions on culling milking cows and for breeding genetically improved stock. High labour costs have led to much mechanisation, such as machine milking, forage conservation and feeding stock, while cows grazing at pasture are able to harvest their own forages more efficiently than can farmers. Low population pressures, hence relatively cheap land, have allowed farms in Western countries to expand in both size and cow numbers.

Unfortunately, this has not been the case for many SHD farmers in most Asian countries. Being in the tropics, feed quality suffers from high temperatures, humidities and often strongly seasonal rainfall patterns. Dairy cows are temperate animals with thermo-neutral (comfort) zones closer to 10°C than to 30°C. Furthermore, high humidities reduce feed intakes which exaggerate the adverse effects of high fibre forages on appetite. A good measure of heat stress, the Temperature Humidity Index (THI), shows milking cows in the lowlands of the humid tropics to be in the 'high stress' and 'reduced performance' zones for much of most days throughout the year. Many dairy specialists correctly argue that potentially high performance dairy breeds, such as Friesians, may not necessarily be the best cattle genotype for tropical regions, except in highland areas or those with low humidities.

There are many socio-economic reasons why the efficiency of SHD farming in Asia has not greatly improved over the last two decades. Granted, numbers of cows have greatly increased in most Asian countries, largely through government support for social welfare and rural development programs. The increased demand for milk (accentuated through school milk programs) and the concept of national food security are the driving forces behind most dairy development initiatives. However, in terms of feed inputs per kg of milk produced or farm milk outputs, improvements have been slow. This is demonstrated by the inability of virtually all dairy industries in SE Asia to markedly improve their self-sufficiencies in milk over the last 10 years (see Table 1.1 in Chapter 1), hence reduce their reliance on imported dairy products.

In addition to the above biological constraints, the other major problem to achieving national dairy development production targets has unfortunately been a common human failing, namely an inability to properly plan for such initiatives, in the short as well as the long term. This chapter discusses this problem at three levels:

first, at a regional dairy program level; second, a 'greenfield' or new farm development level; and third, troubleshooting existing dairy farms that are not performing, even to expectations. Many regional dairy development projects also involve the construction of a series of medium to large-scale dairy farms (say from 200 to 1000 milking cows) frequently on a 'greenfield' site or one with little existing dairy infrastructure. The third level occurs all too often when poor planning has resulted in new or existing farms that do not achieve realistic production and profit targets.

The importance of long-term planning is paramount in any dairy development program. We often hear the comment 'Failing to plan is planning to fail'. Unfortunately, this applies to much of the dairy development around tropical SE Asia.

12.1 Steps in planning dairy development programs

12.1.1 Regional dairy programs

A common problem with many regional dairy development programs is the desire to introduce the stock long before the infrastructure has been fully prepared to support them. Importing pregnant dairy heifers provides a small window of opportunity for their eventual calving and milk production. However, all too often this window is too small to prepare for a heifer's change to a lactating cow, requiring optimum feeding and herd management to settle into their new, often more hostile, environment.

Figure 12.1 lists 10 steps that should be followed in any large-scale regional dairy development program while Figure 12.2 summarises the protocols involved. It is essential to organise markets, milk processors, physical and social infrastructure before introducing stock. The actual cost of milk production cannot be determined until the stock are on-site and their actual, rather than their predicted, levels of performance and required inputs can be quantified.

An additional step that overrides the success of all those in Figure 12.1 is a planned and ongoing supply of finances to ensure each step actually occurs 'on time and on budget'. This requires a long-term commitment from financiers well before the program starts. This budget must incorporate realistic levels of cow performance based on local information or estimates and not those from other often over-optimistic ones, generally derived from temperate, and hence less stressful, environments. The budget obviously needs to incorporate a cash flow as well as long-term loan repayments and should not plan for any profits for several years into the programmed development.

12.1.2 A 'greenfield' or new farm development site

Converting a greenfield site into a profitable and sustainable dairy farm also requires careful planning. The steps to take are similar to those in Figure 12.1

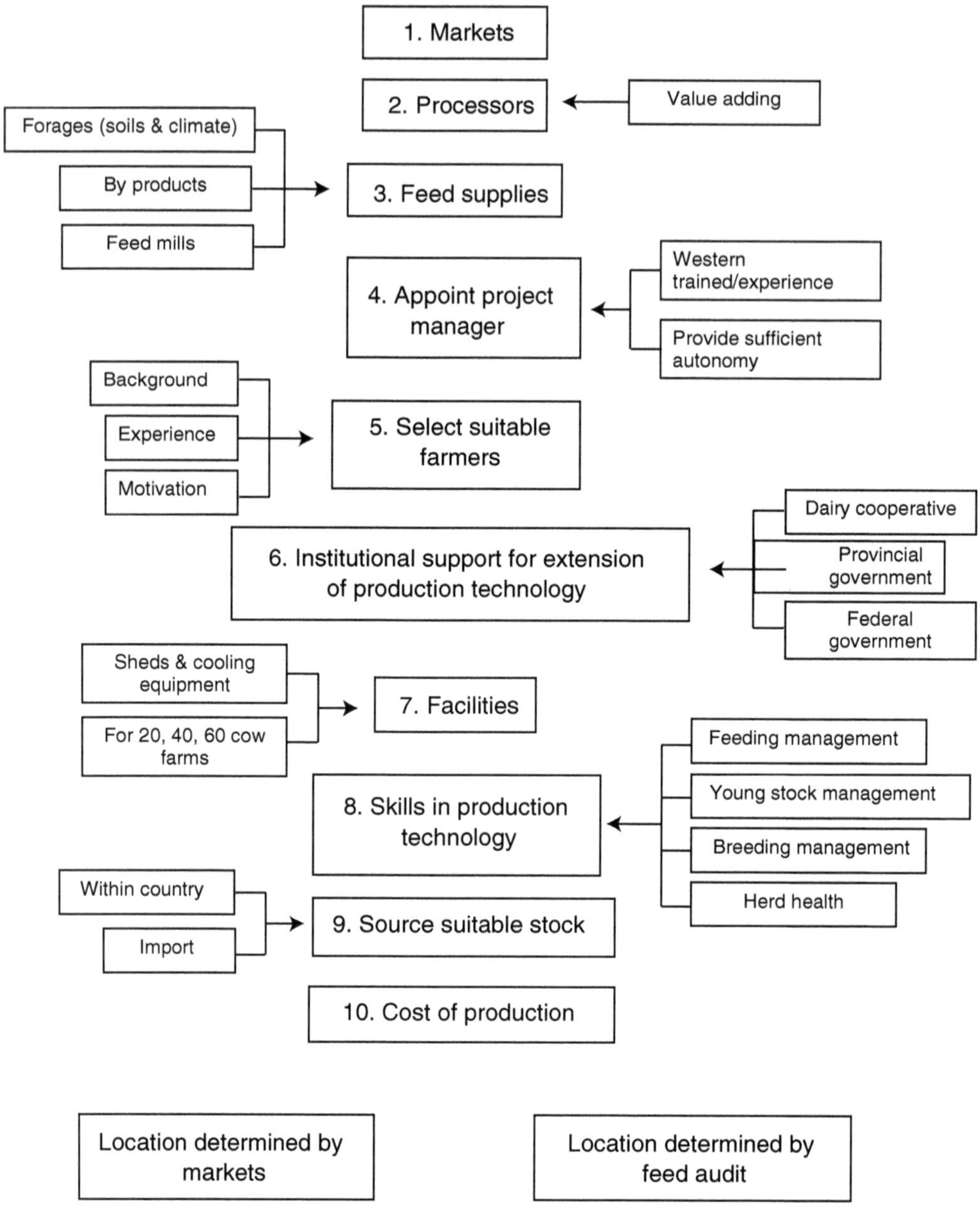

Figure 12.1: Ten steps to be followed in any regional dairy development program.

except that several of them would be taken for granted. For example, one would assume that there is an existing market and milk processors (Steps 1 and 2) or at least one that will definitely develop in time to utilise the raw milk from the new farm. Ensuring a sustainable feed supply (Step 3) and suitable staff (both managerial

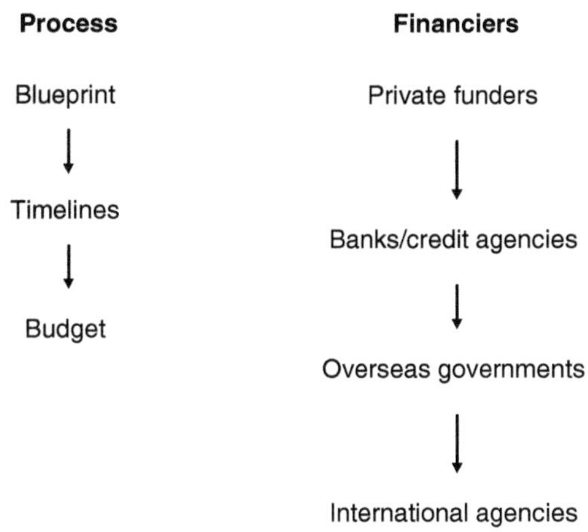

Figure 12.2: Summarising the protocol for regional dairy development.

and general farm staff, Steps 4 and 5) are essential before introducing the stock. The diversity of skills required by dairy farmers and their farm staff has previously been discussed in Chapter 6.

Step 6, the importance of training the farm staff, should be taken for granted as that would have been ascertained before the project started. The basic facilities must be constructed before the stock arrive. Of greatest importance, the assurance of sufficient and timely supply of finances is essential to ensure the project does not stall at any step along the way.

Probably the most important decision that needs to be made for a greenfield site is the proposed stocking capacity, that is the number of milking cow units to be maintained per ha of forage production area. One milking cow unit is one adult cow plus 20% of its replacement heifer, that is assuming a 20% replacement rate per annum. The farm should aim to supply as much of the annual forage requirements as possible, to give the farm management team more control over the supply and quality of such forages than if they have to be purchased off-farm. The hardest part of this decision process is the assumption of annual growth rates of such forages. This has been discussed in detail by Moran (2005), who has concluded that to ensure all forages can be grown on-farm, such target stocking capacities should range from 7 to 10 milking cow units per ha forage production area. Once this has been decided, then a more realistic calculation can be made of the total tonnages of forages that need to be purchased. The annual requirements for the other major dairy feed, namely concentrates (either formulated or sourced as raw ingredients), also need to be ascertained so that long-term sources can be assured early in the project.

Net cash flow means an estimate of farm expenses as well as farm income. The majority of cash is generated by the sale of milk. The shape of the lactation curve

(Moran 2005) means that this can only be consistent from month to month with careful planning to source dairy stock at several times during the early years of the project, and not all at once. All too often such greenfield projects fail altogether or require major cash flow revision because all the stock were introduced at the same time.

12.1.3 Hitting the 'white wall'

The above highlights a classic scenario where 'new' farmers enjoy a rapidly increasing cash flow when all cows calve down over a short time frame. The farmer often then increases his cash input, sometimes into lower priority investments, neglecting the most important ones, such as maintaining a high quality (hence high intake) ration as cows approach mid and late lactation and ensuring optimum reproductive performance (using additional fertile bulls rather than depending entirely on artificial insemination and ensuring all field staff develop skills in heat detection). Persistency of milk production (as quantified by the average monthly decline in milk yield from peak yield) is one of the often neglected, key measures of success of a feeding program and has been discussed in Chapter 8 in Section 8.7.5. It should be of the order of 8% or less per month decline from peak rather than the all too common 12% or more (Moran 2005).

Too often herd milk production rapidly decreases as cows move into the less productive phases of their lactation, reduced cash flows follow and the farm's net income declines to such an extent that its long-term viability may be at risk. Such scenarios are rarely made public as national pride can be at stake, hence it is often repeated by new, inexperienced investors in dairy development.

12.1.4 Troubleshooting an existing dairy farm

Troubleshooting the mismanagement of an existing enterprise can be one of the hardest problems because errors in design and construction of facilities, shortfalls in supplies of feeds, particularly forages, and inadequacies in managing the stock may have already introduced constraints on potential cow and farm performance. We will discuss this using a theoretical case study based on an actual situation.

A 150-milking cow free stall barn farm was established using pregnant grade Friesian heifers, all imported at the same time, in a hot humid environment in tropical SE Asia. Insufficient area was allocated to forage production and very few staff had had much experience with intensive tropical dairy farming. Within its first 5 months of operation, milking stock were suffering from severe weight loss, stock (cows and calves) were dying, milk yields fell to average only 7 kg/d, cows were not cycling post partum and there were increasing numbers of new cases of mastitis occurring every month.

Over the following 5 months the farm management, with consultant advice, were prepared to invest in a series of farm improvements that had dramatic beneficial effects on cow and farm performance. Milk yields and body condition increased and the cows started cycling. These farm improvements included:

- sourcing and developing more area for forage production
- ensuring year-round supplies of clean drinking water
- introducing a mixer wagon to allow for blending the ingredients and mechanical feed delivery
- concentrating on ration formulation to balance daily energy, protein and other nutrient requirements
- introducing a fermentable energy and a rumen degradable protein source
- incorporating a small amount of rice straw in the diet to provide physically effective fibre
- formulating lower cost rations to reduce feed costs
- improving newborn calf hygiene and colostrum feeding
- routinely Californian Mastitis Testing cows followed by antibiotic treatment of subclinical mastitis cases and culling chronically infested cows
- purchasing bulls for natural mating, rather than just planning to practise artificial insemination
- installing a water sprinkler system and cooling fans for better climate control
- introducing recording systems, using both notebooks and computer software, to more closely monitor daily management practices
- establishing a computer system to quantify milk income less feed costs and the proportion of feed consumed by non-productive stock each day
- selling off bull calves and cull stock
- initiating regular faecal and blood sampling and vaccination protocols for better disease management
- using pregnancy diagnosis and record keeping for better reproductive management
- importing pregnant stock with a range of expected calving dates
- improving on farm biosecurity.

Over a 12-week period, following these improved management practices:

- the number of cows dying decreased from five per week to zero
- feed intakes increased from 8.4 to 15.0 kg DM/cow/day
- average milk yields increased from less than 8 to more than 14 L/cow/day
- body condition scores improved from 1.5 to 3.5 units (out of 5)
- the farm manager signed the consultant up for a further period
- the owner was seen more often with a smile on his face.

12.1.5 Importing dairy stock from overseas

Very rarely, if at all, can dairy development programs rely on natural increases of heifers to populate the new regions. Calf mortalities are just too high in most SE Asian countries. For example, Moran (2011) reviewed the published data concluding that a range of 15 to 25% pre-weaning and early post-weaning mortality rates would be typical on many tropical dairy farms, in contrast to the 3 to 5%

(a)

(b)

Figure 12.3: Regional development programs often involves importing large numbers of dairy heifers for many small farms. Airfreighting them in is often more logical as this reduces travel stress. (a) Loading crates of heifers onto the freight plane in Sydney, Australia. (b) Each crate contains 7 to 9 heifers, depending on their age and pregnancy status.

considered normal on well-managed dairy farms in developed temperate countries. Such high mortality rates would also be indicative of large numbers of surviving calves that have suffered permanent health problems leading to reduced lifetime performance. Therefore, the only way to source more dairy stock to improve farm and regional milk outputs, assuming they can be fed adequately, is through an active program of importation. Australia and New Zealand seem to be the countries of choice, although Thailand also has an active dairy heifer export market especially to Malaysia and Vietnam (see Figure 12.3).

When considering importations of dairy heifers, there are two major decisions to be made, namely what genotype is the most suitable and what age should they be on arrival. Unfortunately all too often, the first is considered a 'given' by many decision-makers who plan dairy development policies. That is, they must be 'black and white'! If the dairy region is in the highlands (say above 800 to 1000 m above sea level) and/or in a region without extremes in temperature and humidity, this is often the correct decision. However, there seems little point in requesting Friesian heifers out of dams that have produced 5000 L milk per lactation, because it is highly unlikely that the imported heifers or their progeny will be managed well enough to achieve such milk yields, particularly if they have been imported for smallholder farmers. In most cases, any dairy genotype imported from a developed country is likely to be of higher genetic merit than the typical milking cow in tropical Asia.

Koonawootrittriron (2016) reported that from a genetic analysis of Friesians bred within Thailand, purebred cows had higher full lactation milk yields compared to 50% crossbreds (5050 v 4500 L), but they also had 43-day longer calving intervals (471 v 428 days). Because of health and reproductive problems, he recommended that farmers needed to keep the breed composition of their cattle at ~90% Friesian or lower for Thailand's SHD farms.

Jerseys or their crosses should be seriously considered in tropical dairy systems when climate constraints are apparent and/or when feeding and herd management is very much sub optimum. They are smaller, hence have lower maintenance requirements, have better climatic tolerance (due both to lower milk yields and physical characteristics such as sweat gland density and skin colour) and often better reproductive performance. In areas where financial incentives encourage farmers to produce milk with higher solids content, Jerseys even outperform Friesians. Data from Bangladesh SHD farms, as reported by Moran and Doyle (2015) confirm these statements. There are other dairy breeds that seem to perform better than Friesians in the torrid tropics such as Brown Swiss, or synthetic breeds such as Australian Friesian Sahiwals or the Girolando (from Brazil), while the purebred Sahiwal (from Pakistan) justifies further consideration.

The other decision to make is whether to import pregnant heifers or yearling (virgin) heifers. Pregnant heifers are the most favoured because farmers get 'two for the price of one', assuming the foetus is a dairy genotype. In addition, as the heifer is pregnant (at least diagnosed as pregnant) she does not have to be mated soon on arrival at her new home where there is no guarantee that she will easily conceive. However, with only a few months to adapt to her new environment, there is also no guarantee that the heifer will remain a long-term member of the milking herd once she calves down. All too often one hears stories of very high numbers (up to 30%) of imported heifers being culled and destined for slaughter after having had only one calf. The most likely reason is that their poor feeding management post-calving and their higher genetic propensity to utilise body reserves to produce milk, have combined to result in anoestrus for many, many months post-calving. Such animals have become very expensive dairy beef animals. In the long run, yearling heifers may be better economic propositions than pregnant heifers.

Lastly, and of equal importance when importing dairy stock into tropical Asia, are issues of animal health. All countries, both importing and exporting countries have disease management protocols. Such protocols must be strictly enforced and regularly reviewed, in case of new disease outbreaks in countries of origin. Foot and mouth diseases and brucellosis are the two most commonly talked about but there are others to consider. In one recent example, two diseases, namely bovine viral diarrhoea (BVD) and infectious bovine rhinotracheitis (IBR) were isolated in virtually every aborted foetus arising from one importation of pregnant heifers. These can have long-lasting adverse impacts on cow performance so require additional surveillance to ensure they do not enter the country with the consignment.

12.2 The overarching principles of regional SHD development

Dairy development is generally associated with technical changes to improve milk yield per cow. However, it should be noted that:

- The use of exotic cattle is a rapid and potentially sustainable path to higher productivity, even for small-scale resource-poor farmers and in warm, semi arid or humid climates. However, there have been many repeated failures of such schemes for obvious but often ignored reasons.
- National and local breeding strategies need to address the realities of climate and disease risk to increase the likelihood of successful crossbreeding programs.
- Fodder production and preservation technology should be an integral part of any dairy development program, particularly if it incorporates importation of stock of high genetic merit.
- The success or otherwise of intensive fodder production schemes is more likely to depend on availability of cheap labour, scarcity of land and good access to milk markets than it is on the agro-climatic setting. Where labour is scarce, intensive fodder cultivation practices and the feeding of crop residues to cattle are unlikely to be taken up unless mechanised. Promotion of such schemes should pay very close attention to labour opportunity costs.

12.2.1 Requisites for long-term sustainable dairy development

In summary, for regional dairy development to become sustainable, there must be:

- adequate infrastructure and marketing opportunities
- access to reliable markets for increased milk production
- promotion through government policy
- availability of credit for purchasing of livestock and planting pastures
- available productive and adapted forage species
- ready access to information
- farm management systems that ensure adequate feed throughput the year
- management of animal wastes
- disease control measures including biosecurity in newly acquired stock
- adequate hygiene for milk collection.

12.2.2 How do various smallholder producer models perform?

From their assessment of dairy development in nine different Asian dairy industries, Dugdill and Morgan (2008) summarised a series of 'lessons learnt'. These were as follows:

- Centrally planned models or those where government intervened in milk pricing did not fare well in the long term.
- Government-owned dairies, especially large-scale ones where civil servants managed the business, did not fare well, although there were exceptions such as Chinese centrally owned but market-orientated companies.
- Most of the successful models were private sector-based, as they have more flexibility and are less constrained by regulations than other producer models, such as cooperatives.
- There were concerns, such as if the private sector wanted to maximise profits and reduce risks by using cheap imported (subsidised) dairy commodities rather than set up more difficult to manage, local milk procurement schemes.

  The key lessons for the private sector were then:

- It should become engaged sooner rather than later in the development process. Creative and carefully thought out linkages between smallholder groups and the private sector, such as technical assistance and financial support, will enable smallholders to more easily move up the marketing chain.
- Milk quality, diversity of dairy products and attractive product branding and presentation are prerequisites for persuading modern urban consumers to switch from imports to locally produced milk.
- Value-adding activities can enhance returns to dairying. Selected smallholders close to formal and informal markets should produce high value-added ready-to-drink indigenous and niche products.

The key lessons applying across the entire dairy sector were then:

- Smallholders need accessible and affordable complete packages of technical support services (such as animal health and AI services) to produce milk competitively. Not surprisingly, the key technical constraints are lack of feed and fodder, dairy breeding stock and training. Technical knowhow and skills can be delivered through vocational and outreach training by industry institutions or SHD groups.
- Pro-poor social programs, including school milk programs, need to be carefully targeted and are usually sustainable only if linked to remunerative markets.
- Lactose intolerance is basically a myth because many people seen as non-milk drinkers are increasing consumption of ready-to-drink processed and cultured milks. School milk programs help develop the milk drinking habit while promoting future demand, but should be based on locally produced rather than imported milk.

They concluded with the five overarching principles of SHD development:

1. Smallholder dairying is straightforward in concept but complex in execution.
2. Dairy farmers must be competitive to access markets, by producing top quality milk at affordable prices. Success requires adoption of a complete cow-to-consumer strategy and intervening at every stage of the dairy food chain to ensure profitable product integrity.
3. Strategies for and including smallholders require deliberate and creative development processes that are sensitive to the impact of policies, programs and activities of the farmers themselves.
4. The impacts of such policies, programs and activities on the farmers depend on the local context and the people involved.
5. The private sector must be fully engaged with both government and farmers.

12.2.3 Ensuring a future for smallholder dairying

With growing political attention to narrowing current income disparities, there are many opportunities to use dairying as an instrument for development more successfully. When planning regional strategies, Young (2008) made some salient comments. These include:

- **Think globally but act locally.** As part of a global industry, smallholder dairying has to be competitive in a world of declining trade barriers. The 'one size fits all' approach does not work given the vastly different scenarios across Asia.
- **Focus on the long term**. It is too easy to be distracted by current issues and overlook the long-term requisites that are fundamental for their success.

- **Consider a wide range of possible scenarios.** There are many critical uncertainties, such as climate change and changes in consumer preferences and trade policies. These must be factored in when formulating strategies for SHDs to remain relevant and flexible enough across the region.
- **Design for impact.** The regional strategy is about people, the millions of rural poor, and not about milk or cows. Policymakers must always remember who the intended beneficiaries are and how their lives will be affected by any proposed interventions.
- **Avoid over-prescriptive blueprints.** By developing a range of options, individual countries and regions can select those most relevant to their needs and available resources.
- **Don't overlook financing needs.** Dairy development is capital intensive compared to other forms of livestock production. Finances should be mainly 'up front' because ongoing donor support may not be sustainable.
- **Consider the enabling environment.** Although such strategies can be successful under a wide range of situations, they must take into account the key features of the region, such as its transport infrastructure, electricity and water supplies, financial services, telecommunications, current animal husbandry practices, technical support and market linkages.
- **Adopt a demand-led approach.** The engine of these strategies is growing consumer demand. However, as consumers are better informed, they become more discriminating in their tastes and more demanding of quality, product range and convenience. Without consistency of quality, smallholders may be sidelined for industrial-scale dairy operations or imported product. There may be market niches for smallholder products.
- **Establish preconditions.** Dairy development will not work without social stability, adequate governance and sound macro-economic policies. It must also meet the conditions of political feasibility, administrative capacity and financial affordability.

If the world is committed to reducing poverty and achieving sustainable growth, the future must include more sustainable livestock farming. As there are no 'magic bullets', this requires broad consultations at the country level to customise agendas and define implementation strategies. It also requires having livestock farming work in concert with other sectors at the local national and global levels. It requires building the capacity of smallholders and their organisations, private agribusiness and the state. It requires institutions to help agriculture serve development and technologies for sustainable Natural Resource Management (NRM). It also requires mobilising political support, skills and resources. However, it is difficult for developing countries to strive for sustainability when so many of its population are faced with poverty and hunger. It is difficult to stop people from cutting and burning the forest, eroding the land or destroying the animals and fishes when they and their families are hungry. It is difficult to take an altruistic viewpoint on reducing global greenhouse gas production when there is such a discrepancy between the world's rich and the world's poor. In other words, it is not easy to be green when you are in the red! The need to strengthen moral and religious beliefs and the idea of sacrifice for the common good and survival of future generations are admirable philosophies that should be practised by every element of society, not just those who demonstrate such beliefs in closed communities as part of their everyday life.

Smallholder farming systems, though ecologically sustainable, are basically traditional ones which need further technical improvements to increase the farm outputs, hence profitability, to satisfy the food requirements of rising human population. However, research activities directly relevant to the development of small-scale production systems have received little attention from both scientists and policymakers in the developing countries. Chantalakhana and Skunmun (2002) argue that most scientists and high-level administrators have been overly influenced by, and even overwhelmed with, the commercial production of commodities for export to earn foreign exchange, with insufficient emphasis on import replacement.

Since most SHD farms have the potential to be technically, socially and ecologically sound, and they constitute the majority of milk production systems in tropical Asia, it is not only economical but even ethical, to give real attention and effective support to their sustainable development. Short-term gain and quick profit incentives must be seriously and carefully evaluated against any loss of natural resources and human survival in the long term. This does not mean that, with appropriate management and sound government policies, large commercial dairy farms cannot contribute to the milk flow. However, even after 20 or 30 years of bitter experience, there are still far too many instances where large dairy operations have failed because not enough attention has been given to the basics.

Unlike crop science, most dairy production and health research are confined to the research station or laboratory. Such research can be criticised for not being relevant or practical enough to solve the 'real farm problems'. Unfortunately, far too many dairy scientists are 'compartmentalised' in their scientific thinking rather than intuitively look 'at the big picture' to employ farming systems or a holistic approach to their work. Very few seem interested in or have a real understanding of the concept of sustainable farming and there is little evidence of local farmer participation right from the beginning to the very end of any project. Local farmers' knowledge and problems must form the basis of their research. Changes in technical efficiencies can have beneficial carryover effects on socio-economics. Because village animal production involves the entire family (men, women, old and young – including children), it must consider the needs and limitations of all these potential stakeholders on any one SHD farm.

In addition, research organisations in developing countries need to review their philosophies on science and technology and direct them more to the needs of farmers and local communities, not towards international publications or their own self-interest. As Chantalakhana and Skunmun (2002) express it so eloquently, animal scientists working on sustainable livestock production must be those with the utmost interest in people, and not only money.

In summary, dairy farming would have to be one of the most sophisticated forms of livestock production in the world and regional development programs should only be undertaken after careful and logical planning and with expectations of a long-term investment before profits accrue. To be successful, such programs must involve sourcing, or at least seeking support from, personnel with proven tropical experience in both dairy farming practices and dairy farm business management. Much can be learnt from the litany of failed dairy development projects throughout tropical Asia so these mistakes will not be repeated, as unfortunately occurs all too often.

12.3 The Dairy Smallholder Innovative Program

The self-sufficiency of milk production of many SE Asian countries is very low and in several cases is actually decreasing (see Table 1.1 in Chapter 1). Individual farm productivity is limited by many constraints to increasing per cow milk yields both per lactation and per lifetime. These include:

- 10 kg/day milk, compared to a realistic 15 kg/cow/day
- 15 to 18 months calving intervals, instead of 12 to 15 months
- 30 to 33 months age at first calving, rather than 24 to 27 months
- high numbers of cows suffering from climatic stress, as is evident by their respiration rates exceeding 70 breaths per minute
- high levels of subclinical mastitis
- unacceptable levels of cow lameness
- high mortality rates in young stock, often with little diagnoses of the cause
- high wastage rates within milking herds due to poor feeding and herd management.

The Profitable Dairy System's (PDS) project team has developed a strategy to work closely with SHD farmers and their service providers throughout SE Asia through a new capacity building program, entitled the Dairy Smallholders Innovative Program (DSIP). Its major objective is to develop individual farmer skills primarily in dairy production technology to improve the productivity of their milking herds to stimulate their farm performance and profitability, hence long-term sustainability.

The DSIP incorporates five closely related strategies:

1. Crystallising the major farm activities into a series of '**Golden Rules**'.
2. Quantifying the beneficial impacts of adopting these Golden Rules using **Key Performance Indicators**.
3. Demonstrating them on a series of **model farms**, which is more likely to encourage farmers to adopt them than by other capacity building approaches such as classroom-type seminars.
4. Disseminating the benefits of improved farm management practices to a large numbers of SHD farmers through a series of structured **discussion groups**.
5. Providing each participating farmer with the opportunity to more closely monitor his/her own farm productivity and economic performance using a **Farm Business Management** approach.

Many of the SHD industries in SE Asia have been hampered by the low relative performance (milk production and herd fertility) of their dairy stock, particularly in recent years. Much of this is due to the poor management skills of the farmers to improve their levels of dairy production technology. The DSIP has been specifically developed to provide these farmers with an adult learning program based on:

- a series of simple 'take home' messages (namely the Golden Rules)
- feedback mechanisms to assess their comprehension (namely Key Performance Indicators)
- 'hands-on' experiences via practical workshops and demonstrations on model farms
- firsthand validation in collaboration with objective measures of changes in farm business performance (via the model farms)
- facilitated sharing of knowledge (within farmer discussion groups)
- an opportunity to assess on-farm changes (improvements) in farm profitability and sustainability through individual monitoring of their own farm business performance.

12.3.1 The Golden Rules

These together with the associated program on Producing Quality Forages, have been outlined in Section 7.3 of Chapter 7.

12.3.2 Key Performance Indicators

These are essential to quantify the initial farmer performance, then to provide realistic targets for the life of the Program. These have also been outlined in Sections 7.4 and 7.6 of Chapter 7.

12.3.3 Model farms

There is a close relationship between animal welfare, cow performance and farm profitability on SHD farms. Similarly, there is a general lack of awareness of this association by poorly resourced tropical farmers. Farmer-orientated capacity building programs are then a high priority.

Changes in a single farm practice can have a diversity of outcomes. Researchers use traditional approaches to scientific logic in a research environment, to try and hold all other variables constant when assessing the direct impact of one particular variable. However, farmers live in the 'real world of commercial agriculture' where any single variable can rarely be held constant on-farm. Therefore, capacity building for poorly resourced and skilled farmers ideally needs to be undertaken on-farm.

The model or demonstration farm highlights the dynamic nature of farming as it can expose farmers to practical innovations and new ideas. There are various ways in which such innovations can be introduced to a group of farmers, but the important objective is for these farmers to understand how it impacts on 'the bottom line', that is the herd performance and farm profitability. In addition, animal welfare is a very complex part of farm management which unlike other strategies, say feeding a better quality ration to milking cows to increase milk yields, is likely to impact on many aspects of herd performance when practice changes are made. In addition, animal welfare is directed to satisfy or improve the animals' coping mechanisms, with the success or otherwise being made apparent through a variety of performance parameters.

Farmers are experiential learners in that they learn more by doing something themselves (and being able to monitor its impact) rather than by the more traditional learning programs of classroom tuition and short 'hands-on' practical sessions. However, classroom sessions are important because they provide the opportunity to explain the theories behind such practices, which are an integral component of any learning process. Better comprehension of 'why things happen' will improve the understanding of 'how to make things happen', because all too frequently 'things do not go according to plan' because of various unknown and/or unexpected consequences of farmers' actions.

Since farmers learn more by watching and then doing, they need to be provided with every opportunity to watch. This can be provided on the farms of collaborating farmers or, better still, on a model farm. The latter is more desirable because there is more control over farm activities and it is easier to monitor the impacts of changes in farm practices. Using selected farmers who agree to allow their farms to be more closely monitored will provide a control situation so farmers can more easily see and understand the impact of any direct changes in herd performance and farm profitability as a result of such improved management practices.

Model farms should then be established to provide farmers and service providers with an overview of the cause and effect of modifications in farm practices. This is ideally suited to demonstrate the impacts of improved animal welfare. For example, the many benefits rising from the construction of a more 'cow friendly' shed can be demonstrated by comparing herd performance with that of an existing more traditional shed construction. As SHD farmers in many countries routinely travel twice each day to deliver their milk to milk-collection centres, establishing a model farm in close proximity to the collection centre should encourage them to frequently visit it to monitor progress in any demonstration trial. Offering regular field days, at which the latest results are discussed, would further help in their dissemination. Although such a model farm may be the initiative of the public service providers, if well-managed and effective, it is likely to create interest among the private sector which could then be levered to provide additional resources for its operation.

The project team will need to select model farmers based on their credibility in the local dairy industry, their motivation to want to improve their farm practices and to share their knowledge. One motivation for an already efficient farmer to become a model farmer would be the offer of a full business analysis of his dairy operation to help him make more meaningful decisions for future development. There are many low investment modifications to existing farm management that can be demonstrated on a model farm. In fact, Moran *et al.* (2016) have categorised beneficial changes in traditional SHD farm management, noting that many of these only require improving the time management rather than actually having to invest actual finances. In other words, many changes in management practices have negligible costs apart from spending more time on any one task.

The modus operandi of model farms

The diversity of dairy production systems means that there should be more than one type of model farm. They could vary in herd size, housing and milking parlour facilities and source of feed supplies. A series of model farms should be developed, so farmers could see firsthand how improved farm management provides long-term benefits to their SHD industry. Such a farm requires careful planning, preparation and construction to ensure it has a sustainable future as a farmer extension tool. The following is a list of some of the prerequisites:

- These farms should be located in a dairying region, close to other commercial dairy enterprises.
- They should be of a commercial size and relevance to local dairy farmers within the next 5 to 10 years, for example, they could be between 20 and 40 milking cows (and associated young stock). If they are too large, farmers will have difficulty relating to them, hence not willing to adopt any improved management practices. The size of the forage production area should be such as to provide fresh forage for say 8 to 10 milking cows per ha forage per year. This may or may not include home-grown forages conserved as silage.
- The facilities should be constructed of materials that are readily available and likely to be used by other dairy farmers.
- They could be stocked with both locally sourced and imported dairy heifers as these will become the future benchmark of the industry.

- They should be managed by experienced dairy farmers with proven track records of innovative dairy farming.
- The day-to-day management would be the responsibility of the owner/manager but he will be guided by an advisory group of experienced dairy specialists, representing a range of agencies, from the investors, to expatriate dairy advisers to government advisers. This group will meet several times each year to monitor progress in each farm's economic performance as well as their role as extension tools.
- They would be operated as commercial businesses with additional resources available to develop effective extension activities, such as regular visits from local dairy farmers and other dairy stakeholders.
- The milk produced would be sold through the normal outlets, either through a milk collection centre or direct to an established milk processor. Consideration could be given at a later date to value-add some of the milk.
- In addition to the initial finances to purchase (or rent) the land and construct the facilities, agribusiness and maybe even government or other dairy industry stakeholders could be invited to invest. These investors will then be invited to become involved in the overall management of the farm.
- As well as documenting cow and farm performance, all data on cash inputs and outputs would be collated and regularly summarised to form integral parts of the extension messages.
- There should be regular meetings (small groups of farmers visiting) and an annual regional forum to extend the correct farm management practices.

12.3.4 Farmer discussion groups

As already discussed, farmers are motivated to improve their management practices more by seeing than by reading or attending training programs. They are also not good at keeping records unless they are sufficiently motivated.

Already successful in most developed dairy industries, for some reason discussion groups have rarely been utilised in tropical Asia. Granted there has to be a set of ground rules for any successful discussion group, but these should not be too onerous for the concept to be more fully accepted in Asia.

Farmers generally like gathering together to share ideas:

- This is even better if they can visit a farm and see how the management is and learn from what the farmer does well.
- The model farm is an ideal venue for such a visit.
- Groups can encourage innovation and new ideas as people are more creative in groups.
- Groups allow ideas to be more effectively communicated and applied.
- Groups can enhance networking and learning through pooling of skills and knowledge from individuals.

- Such groups must have strong aims or direction and clear objectives and aims must meet people's needs.
- These objectives need to be regularly assessed to see if they still meet people's needs and expectations.
- The group should collectively plan the program for the next few months.

One good way of getting farmers to express what they want to gain from being part of a discussion group would be to undertake a 'brainstorming' session to answer the following questions:

- What farm management issues are of most concern to the group? Develop a list of say five to eight of them.
- How should the group tackle each issue in the next 3 to 6 to 12 months? The group should then set achievable goals.
- What are the group's short and long-term priorities and how shall they set realistic and achievable goals?

The group needs a strong facilitator. This can be a farmer or a service provider who keeps all members regularly informed about future plans. Early meetings should have a sense of purpose. If nothing else, members should leave with a sense of achievement and a feeling that the group is moving towards worthwhile goals.

The facilitator should:

- gain the trust and confidence of the group members
- help the model farmer collect, analyse and interpret the farm data to be discussed
- ensure the model farmer is fully prepared for each group meeting
- facilitate the meeting, encouraging all farmers to contribute their ideas
- keep the group on target to discuss the key topic for the day
- seek resource people to stimulate further discussions
- summarise the day's discussion and major conclusions
- ensure the group sets the agenda for the next meeting.

Success of a group depends on several things, but includes:

- appropriate group size for purpose
- clear focus, aim or objective
- revisiting goals of the group often and measuring progress
- developing clear boundaries and ground rules
- developing clear expectations, roles and responsibilities
- ensuring participants have a degree of enthusiasm/willingness
- ensuring enjoyable social activities for a job well done
- having good interpersonal relationships
- ensuring individual's needs are met

- ensuring relevance of the chosen issue or topic
- ensuring resources are available (participants' fees or group funding)
- ensuring there is a skilled facilitator and also support staff
- providing skills training is appropriate for participants' role in the group
- respecting other people's views and opinions.

In summary, discussion groups are most successful if:

- They are well organised and have a planned approach.
- They have well-planned activities with a clear sense of purpose.
- There is strong 'ownership' of the group's direction by the farmer members.
- Groups are effectively facilitated.
- Information and planning is driven by regular evaluation of the group and its achievements.
- The facilitator and host farmer have a good idea of the key drivers to success of the business and knowledge of potential barriers to success.
- The host farmer is always comfortable with what is being presented to the group.
- The session concludes with a summary of the discussion and any opinions or thoughts that have been raised, to allow the group to clarify points of discussion and see that the discussion is targeted towards improvement.
- The group members can develop a good insight into what the model farmer is doing particularly well, so they can take home potential 'successes' to apply to their own farm business.

12.3.5 Monitoring business performance on participating farms

It is possible to quantify the cash flow, cost of milk production and business performance on any SHD farm quite easily by asking a series of specific targeted questions. Such an exercise on each DSIP participating farm will provide individual baseline data from which to predict the likely financial benefits in achieving target KPIs. More importantly, it will also provide comparative data on the changes in farm performance of model farmers.

Assessing farm business performance is the topic of Chapter 11 with a questionnaire presented in Appendix 3.

13

The final words

This chapter highlights some of the developments and on-farm applications still required for tropical dairy farming to become and remain more efficient and profitable. It also provides details on an online course on recommended feeding management of tropical dairy cows and shows readers how to access it.

13.1 Constraints to dairy farm developments in Malaysia; a national case study

The following discussion refers specifically to Malaysia but it would apply equally to the dairy industries of many other Asian countries. Moran and Brouwer (2013a, 2013b) undertook a survey of 30 dairy farms in Peninsular Malaysia, the data of which has already been summarised in Chapter 11 in Section 11.3. Of equal interest, they also developed a list of dairy farm production constraints to Malaysia's milk industry, these being:

- **Shortages of land** for growing forages means that farmers must greatly depend on purchased supplies of forage over which they have less control of feed quality.
- **High feed costs.** Dairy cattle pellets and molasses, the most commonly used energy feeds, have the highest cost per unit energy of other alternatives.
- **Shortages of 'quality cows'.** All too often farmers blame the cows, rather than their lack of management skills, for poor stock and herd performance.
- **Poor infrastructure**, which includes power, water and local labour supplies.

- **Poor support from service providers**. This becomes increasingly apparent since many of the government services are free and this provides little incentive for the private sector to develop and compete for such services.
- **Poor reproductive performance**. Reduced milk yields through under-nutrition may be accepted as 'normal' by many farmers, but in many SE Asian countries farmers routinely complain about high ages at first calving, lengthy calving intervals and poor conception rates to AI.

In addition to those constraints highlighted in the farmer survey, they noted the need to improve:

- **Climate control**, through installing fans in every shed, sprinkler systems in most of them and higher roofs with open air spaces at the apex.
- **Cow comfort**, through thick rubber mats, free stalls, better shed hygiene and closer attention to treat and prevent lameness.
- **Adequate supplies of clean drinking water.**
- **Better agronomic management of existing forages**, such as routine use of inorganic fertilisers and shorter harvest intervals, to reduce forage fibre levels, stimulate more rapid feed digestion rates and so promote better appetites and cow performance.
- **Improved milking hygiene practices** such as more frequent replacement of rubber liners and greater attention to mastitis management.
- **Financial assistance** to source more feed to better feed existing animals before sourcing credit to buy more stock.
- **Capacity building** (both advisers and famers) with the two priorities being first, nutrition and feeding management and second, farm business management.

Despite the fact that Malaysia's farm gate milk prices are among the highest in SE Asia as is their milk to feed price ratio (Ahuja and Staal 2013), domestic milk production in Malaysia has stagnated over the last decade. There are many contributory factors for Malaysia to have 'flatlined' in domestic milk production. Among the possible reasons are:

- Dairy farming is expensive in the tropics because of lack of suitable stock, forages and quality ingredients for concentrate formulations. Farmers react by demanding higher milk prices but processors respond by purchasing more of their dairy ingredients from cheaper overseas sources.
- The poor understanding of most dairy farmers of how dairy cows need to 'function' to become profitable units of investment. They are just not simply 'cash cows'. They are living animals that require a degree of nurturing to remain healthy and productive. Basically dairy cows are temperate animals, so to fully utilise their propensity to produce milk they need to be provided with as comfortable an environment (in a very torrid hot and humid climate) as possible.

- Many farmers enter the industry with little, if any, practical experience in managing a dairy enterprise. Ideally, novice dairy farmers should have spent time either growing up on a farm or at least working on one, before they decide to invest their assets and livelihood in a dairy farm.
- Unfortunately, the banks and other credit agencies often lack the technical skills to assess the suitability of an applicant for a loan to invest (and further invest) in dairy farming. Consequently, unwise dairy loans can and have been made.
- On the whole, few farmers are business minded as they only consider profits through increasing unit milk returns rather than taking a more logical approach, namely reducing unit costs of milk production. This can be done either through reducing unit prices of farm inputs or diluting many of them through increased levels of per cow milk production.
- Over the last decade, there have been several attempts to establish large-scale intensive dairy enterprises in Malaysia. Unfortunately, most of these have failed to produce a long-term, sustainable and profitable dairy feedlot in the country. In other tropical Asian countries successful models are apparent, so one should expect at least one such enterprise to have evolved within Malaysia.
- There seems to be a lack of structured and ongoing encouragement within government advisers for farmers to share their problems and discuss possible solutions in dairy production technology. Other countries have dairy cooperatives in which communication lines are better developed between farmers. Farmers can learn a lot from each other rather than having to rely on government advisers, many of whom lack the practical experience in the day-to-day activities of a tropical dairy enterprise.
- There has been one very recent example of a group of dairy farmers forming an association to facilitate discussions of both dairy production technology and national government policies for future dairy development. This is the Association of Dairy Farmers Malaysia, which comprises mainly Indian dairy farmers and stakeholders.
- There are many free services provided by the government. In Malaysia the approach has always been that governments should look after infant industries such as dairy farming. However, a common result occurs – if farmers do not have to pay for such services themselves, they tend to undervalue them.
- Many of the farmers do not use government veterinary services (and only some use private veterinarians) or artificial insemination (they use their own bulls), because they find the services unreliable.
- There appears to be a gross shortage of experienced and competent private dairy farm consultants in Malaysia. The government is becoming a training ground for those who can eventually freelance their skills (advisers work in government first and then go out on their own). Universities should also provide such advisers both from within their teaching ranks and from

well-trained graduates in all aspects of dairy production technology. However, in dairy farming, this has hardly occurred in Malaysia.

- As in most developing dairy industries, much of the technical support originates from trained veterinarians. However, veterinarians are trained to attend only to sick animals and not healthy and potentially highly productive stock such as milking cows. Formal university training in dairy science requires a greater emphasis on the practical applications of the physiology and the economics of increasing per cow, per herd and per farm production. Unfortunately, this is a common problem in many SE Asian veterinary science university curricula.
- Many of the nutrition advisers in the dairy industry were originally trained in monogastric nutrition (poultry and pigs), whereas specialist dairy nutritionists require a theoretical and practical background in ruminant nutrition and physiology.
- In addition to providing advice on nutrition and feeding management, dairy farmers, both large and small, require consultants with practical knowledge in many aspects of dairy production technology. These include forage production, rearing young stock, optimising cow comfort (through appropriate housing and climate control), milk harvesting and, of even greater importance, farm business and entrepreneurial skills in managing a large investment such as a profitable and successful dairy enterprise.
- Government agencies could be more proactive in fostering relationships with the private sector through the formation of both formal and informal Public Private Partnerships. These should be seen as collaborative rather than competitive.

National policymakers in Malaysia's dairy support services, both government and agribusiness, would benefit from addressing much of the above in conjunction with greater collaboration with some of their SE Asian neighbours. The FAO in SE Asia is represented by APHCA (Animal Production and Health Commission for Asia and the Pacific), of which Malaysia is a member. It has recently established the Asia Dairy Network (APHCA 2013) to improve communication and knowledge-sharing among stakeholders in Asia and the Pacific region along the entire dairy value chain. Recent initiatives in this Network include an E-Conference on agro-industrial by-products for smallholder farmers (Moran 2014), a 'Question and Answers' open forum, an E-learning program in feeding management on tropical smallholder dairy (SHD) farms (see Section 13.5 of this chapter), in addition to regular sessions at future Asian international animal science congresses.

13.2 Risk in tropical dairy farming

Any business must live with risk and dairy farming is no exception. Farmers need to understand the types of risk and concentrate their efforts on dealing with those under their control. There are some specific strategies that dairy farmers can undertake and these are discussed in this lesson.

A dairy business is made up of a mix of people, livestock, natural resources, technology, economics and finance. Farmers buy their inputs to production and sell their milk in the markets. What they do is subject to laws and affected by things governments do. Running a farm business means dealing with lots of risks. There is risk about how much grass they can grow, how much milk their cows will produce and of what quality, hence unit price. There is risk about what their farm production will cost and how much it is worth. There is risk affecting how much spare cash they will have to pay interest to the people who have lent funds to their farm business. There is risk about what governments from all over the world might do that will affect how their farm business is able to operate.

In other words, farming is a risky business. Most farm decisions involve risk or uncertainty (Moran 2009a). Risk and uncertainty challenge the technical and managerial abilities of all farmers. They impact on every part of farming, its productivity, stability, resilience and equity. Uncertainty is defined as imperfect knowledge, or not knowing what is going to happen in the future. Decisions that do not have a single outcome are then uncertain decisions. Risk is only present when the uncertain outcomes of a decision are regarded as significant or worth worrying about. Rather than the risk, it is the consequences of the risk that matter because they have to be managed to achieve resilience in the farm business. Risk and uncertainties can create opportunities and rewards. If the risk is known, profits should have already been made.

13.2.1 Types of risk

There are a great many risks in farming, such as:

- business (farm performance) v financial (other people's money)
- external (outside farm) v internal (inside farm)
- short term (this year) v long term (future years)
- good v bad risks.

Tables 13.1 and 13.2 present risks encountered in different management activities on SHD farms.

Table 13.1. Risk management strategies available to smallholder dairy farmers.

Management activity	Strategy	Type of risk reduced
Production	Use of stable enterprise - dairy genotypes	Yield, technology, policy
	Use of proven forages - e.g. Napier grass	
	Diversifying enterprises - in livestock - in cash crops - within seasons - across farm area	
	Maintain flexibility - over time - in durable assets	
	Keep reserves - seed - fodder - concentrate ingredients	
	Use risk-reducing inputs	
	Share-leasing	
	Assess new technology	
	Seek information	
	Develop annual operational plan	
	Develop animal health and disease management plans	
Marketing	Spread sales over time	Price
	Arrange alternative outlet	
	Seek barter opportunities	
	Consider value-adding and niche marketing	
Financial	Maintain high equity ratio	Financial, yield, price
	Maintain credit worthiness	
	Maintain a cash reserve	
	Develop relationship with cooperatives	
	Maintain farm assets	
	Maintain social network	
	Off-farm employment	
	Develop farm business plan	

13.2.2 Managing risk

Risk minimisation is a major goal of farming.

- Managing risk involves planning for the unexpected
- Most on-farm risks over which farmers have greatest influence:
  - have moderate to high likelihood of occurring
  - have some nutritional basis.

Table 13.2. Types of risk based on farmer's ease of influencing.

Ease of influencing	Risk & Likelihood of occurrence (*)	Strategy to rectify risk
Difficult Outside farmer control	Adverse climate (H)	Cannot be overcome
	Poor industry infrastructure (M)	Government policy and investment
	Uncertain industry future in region (?)	Government policy
	Poor industry technical support (M)	Government training programs
Probable Farmer has some influence	Inappropriate breed (M)	Government support
	Low base milk price (M)	Government, agribusiness and cooperative negotiations
	Poor dairy genetic merit (M)	Government importation programs
	Low post-farm gate milk quality (M)	Cooperative infrastructure
Possible Farmer has great influence	Low farm income (M)	Feeding and herd management
	Inadequate water (M)	Farm development
	Poor forage quality (H)	Agronomic practices
	Low forage availability (H)	Farm development
	Poor concentrate quality (M)	Cooperative program
	Low concentrate feeding (M)	Feeding management
	Adverse diseases (M)	Herd management and cooperative programs
	Low farm milk price (M)	Feeding management
	Poor milk composition (M)	Improved feeding
	Poor on-farm milk quality (M)	Improved hygiene
	Inefficient effluent system (M)	Facilities and herd management
	Adverse heat stress (H)	Shed design and herd management
	Poor animal health (H)	Herd management
	Poor reproductive performance (M)	Herd management
	High cost of production (M)	Farmer management skills
	Poor young stock performance (M)	Feeding management
	Low farm profitability (M)	Farm and herd management
	Increased farm investments (M)	Cooperative advice

* M, moderate; H, high.

On-farm risk management

There are some specific ways that dairy farmers can manage on-farm risk. These include:

- minimising the number of non-pregnant milking cows, particularly those non-lactating
- ensuring sufficient herd replacements
- initiating simple recording systems
- developing annual feed budget
- prioritising farm investment, 'feed before breed'

- concentrating their efforts on low cost management strategies that return greatest income
- becoming active member of local milk cooperative
- becoming an information seeker
- seeking professional support before investing in farm infrastructure
- developing appropriate staff training and management system
- implementing herd health program (record and monitor activity systems for health and production).

Every day farmers must make decisions and strategies to handle risk. These should be prioritised into those within and outside the farmer's control.

13.3 The relevance of 'high technology' to tropical dairy farming

Dairy farming is one of the most sophisticated forms of livestock production and researchers are continually developing innovations in dairy production technology and farm management. However, with many of these innovations originating in countries with advanced dairy industries, we need to ask the question about their relevance to SHD farming in the tropics (Moran 2009a).

Over the last 20 years, most Western countries have produced sophisticated dairy production systems:

- herd sizes have grown
- more efficient feeding systems have evolved
- these days farmers routinely test their cows for production, milk quality and mastitis
- farmers use this information to cull cows and select breeding stock
- high labour costs have led to much mechanisation
- cheap land and grazing cows have led to large herds and farms.

But how relevant is this for smallholder Asian farmers? Furthermore, some of these systems have been unwisely transplanted into tropical smallholdings, such as the all too frequent poor feeding management of high genetic cows. All too often, such advanced genetics and technology is considered the panacea for many Asian dairy industries. However, basic dairy husbandry must be addressed before the potential of such 'high tech' innovations can be more fully realised.

13.3.1 High technology versus appropriate technology

The term 'high or new technology' can be defined as the latest production technology in current or potential use by developed Western dairy industries.

Much of this has been developed in advanced temperate countries with far more sophisticated and productive dairy industries and less challenging climatic environments than in Asia. This is likely to be the case for many years to come.

Advances in biotechnology offer potentially large benefits as long as they are combined with other appropriate farm practices. But today's investments are all too often driven by commercial interests, many of which may have limited impacts in Asia.

Rather than 'high tech', Asian dairy industry policymakers should be looking for appropriate technology that:

- is in accord with the farmers' objectives and constraints
- matches the consumers' demands in accessible market outlets
- leads to sustainable livestock operations.

The relevance of such technologies depends on the capacity and objectives of the farmer. But because SHD farmers have low labour costs and few economies of scale, how can they best make use of them particularly considering the fact that up to 30% of potential livestock production can be lost through animal health issues?

Dairy production technology can be broken down to nine key task areas in the supply chain for any dairy farm, no matter its size or location, as summarised in Figure 3.1 in Chapter 3.

Table 13.3 categorise these technologies into:

1. Those currently very relevant to all sized dairy farms in tropical Asia.
2. Those that may be currently relevant to large-scale dairy operations in tropical Asia.
3. Those not likely to be relevant to smallholders within the next 5 to 10 years.

The term 'latest generation' refers to the most recent stage of commercial evolution in particular farm aids.

Table 13.3 is quite subjective and may be incomplete. However, its contents could become the basis of lengthy discussions between dairy production specialists into the future. Whatever their outcome, it is important to differentiate between high technology and appropriate technology.

Therefore, not all new technologies are currently relevant to the needs of tropical SHD farmers. In fact some may be detrimental to farm profits as the cows are not sufficiently well managed to benefit from their inclusion in the farming system (see Figure 13.1). The term 'appropriate technology' is more relevant when assessing new aids to dairy production technology (see Figure 13.2).

Table 13.3. The relevance of 'high technology' to developing tropical dairy industries.

Links in the production supply chain	Very relevant in all sized dairies	Maybe relevant in large dairy farm	Less relevant on SHD farms for next 5 years or so
1. Soils and fodder production	Effluent distribution systems Macro nutrient fertilisers Micro nutrient fertilisers Latest generation pesticides Latest generation herbicides Most farm mechanisation Latest generation plant genetics	Some farm mechanisation Commercial silage additives	Genetically modified plants Organic production technology Biodynamic production technology 'Alternative' fertilisers
2. Young stock	Electrolyte fluid replacers Thermometers to monitor sick calves	Routine antibiotic therapy Immunoglobulin monitoring equipment	Computerised (automatic) calf feeders Unnecessary overuse of antibiotic therapy
3. Nutrition and feeding	Macro mineral supplements Access to computer software to formulate rations Individual animal identification	Computer software to monitor farm costs Computerised animal identification systems Total mixed rations	Probiotics Micro mineral supplements Vitamin/micro mineral injections Computerised feed dispensers
4. Animal health	Latest generation drugs Latest generation vaccines Access to most modern veterinary practices	Computer software to document animal health procedures Access to some very sophisticated veterinary practices Walk-over weighing scales to routinely record changes in live weight	Routine blood profiles Excessive use of antibiotics
5. Reproduction	Artificial insemination technology Aids for heat detection Single sexed semen	Oestrus synchronisation Tail paint, cameras and activity metres to assist with oestrous detection	Electronic oestrous detection

(Continued)

Table 13.3. (Continued)

Links in the production supply chain	Very relevant in all sized dairies	Maybe relevant in large dairy farm	Less relevant on SHD farms for next 5 years or so
6. Breeding	High genetic merit semen	Exotic high genetic merit heifers	Embryo transfers
7. Environment	Effluent management systems	Village biogas systems	–
	Farm biogas systems	Climate controlled sheds	
	Advances in building designs	Solar panels	
	Latest generation heat stress alleviation systems		
8. Milk harvesting	Bucket milking machines	Automatic teat cup removers	Bovine somatotrophin (BST)
	Herringbone dairies	Automatic cow exit	Robotic milkers
		Computerised (information) technology	Rotary dairies
			Automatic laboratory milk assay equipment
			Rapid exit dairy systems
			Computers in the milking parlour for individual cow monitoring
9. Value-adding	Farmer access to equipment to value-add raw milk	Village access to equipment to value-add raw milk	The latest advanced equipment to value-add raw milk
	Processing, packaging and marketing at the dairy coop level		

13.4 'Words of Wisdom' from Asia Dairy Network's E-Conference on by-product feeding

During Nov/Dec 2013, the Asia Dairy Network of the Animal Production and Health Commission for Asia and the Pacific (APHCA), in collaboration with the FAO Animal Production and Health Division, conducted an E-Conference entitled 'The role of agro-industrial and forestry by-products in the feeding of dairy animals in Asia and other tropical regions'. Over 530 people participated in this E-Conference, submitting 288 contributions over the 4-week period (Moran 2014).

Figure 13.1: This animal is a US dairy heifer, imported into a north Vietnamese dairy feedlot, where the climate stress is extreme but onto a farm with no tunnel ventilation barns. How it will perform as a milking cow is unknown. Will it reach the desired 40+ L/day peak milk yield? If so, will it be able to easily conceive soon after calving? Its purchase was a risky decision.

Figure 13.2: Everywhere throughout the tropical world, forage maize is grown to supplement forage supplies on dairy farms, large or small. These bags of maize silage (on a Kenyan smallholder farm) are very simple to make as they are easy to compact and seal. Even though the initial adoption of the technology may be slow, once farmers see the benefits firsthand, it will become routine in their feeding management.

Many of these contributions provided valuable insights into the feeding and management of dairy stock in the tropics and some of these have been selected and collated below as 'Words of Wisdom' from the E-Conference. Together with the authors, these are categorised into three headings below, namely:

- by-product feeding
- feeding and herd management
- technology transfer.

13.4.1 By-product feeding

Treating straw just to improve its digestibility seems to be mostly a waste of time, when we can easily feed urea and sugar to the animal and thereby improve the digestion of straw within the digestive tract. Dr Steve ***Sutherland*** *of Australia*

Plant breeders have paid little attention to the quality of the by-products and not explored whether the quality could be improved without diminishing the quality and quantity of the main product. Dr Bob ***Ørskov*** *of UK*

The failure of technologies like urea-ammonia treatment of straws is not because of any gap in the technology. It is mainly because farmers, first, do not have time to treat their poor quality crop residues as they want ready-made feedstuffs at their doorsteps and second, because of a general lack of awareness of the methodology and potential benefits. Dr M ***Bakshi*** *of India*

Sometimes it helps to forget about classifying and grouping items as by-products, commodities, forages, energy supplements, protein supplements, food wastes etc., but rather just look at them all objectively as ingredients available at the farm gate for a certain price. Local or imported, it doesn't matter, as long as it is available, and the price is known. Then look at the analysis and formulate a ration to the desired specifications. Dr Steve ***Sutherland*** *of Australia*

Instead of costly chemical and physical treatments to render dietary protein undegradable, it would be more appropriate for low to medium milk producing cows to balance the rumen through less expensive locally available feeds for maximum microbial growth. Dr Ghulam ***Habib*** *of Pakistan*

13.4.2 Feeding and herd management

With the majority of the local dairy breeds in south Asia low producers, the question arises 'Do we need bypass protein supplementation for these cows?' Dr Ghulam ***Habib*** *of Pakistan*

For high producing cows (> 15 L/cow/day), intestinal supplies of amino acids from microbial protein alone may not be adequate and will then require undegradable protein supplement to complement their intestinal protein supplies. Dr Ghulam ***Habib*** *of Pakistan*

Straw feeding is suggested for low yielding cows and as a maintenance ration with some supplementation of critical nutrients. For moderate to high yielders, feeding of green fodder/tree leaves (if green grass is not available) is a must. Dr Ali ***Akbar*** *of Bangladesh, Dr Sahoo* ***Biswanth*** *of India*

Despite the numerous extension programs promoting silage as a potential quality forage, SHD farmers do not want to readily adopt it because it requires 'double handling' of forage, just like wilting fresh forage to reduce excess feed water. Once they have manually harvested the forage, most SHD farmers simply want to feed it to their stock. Dr John ***Moran****, Australia*

Farmers find it easier, time and labour saving and had the idea that green fodders are more nutritious than silage. Therefore, farmers are generally more interested in feeding green fodder rather than making silage and only when they will produce it in large quantities and much more than the requirement, they will be interested. Dr M. ***Akbar*** *of Bangladesh*

Water is almost always a neglected item in many smallholder dairy farms where all too often, milking cows are tied up all day with little or only occasional access to drinking water. Dr Sharif ***Chowdhury*** *of Bangladesh*

Excessive consumption of soluble proteins fail[s] to increase milk production, adversely affects reproductive performance and cause[s] excessive excretion of N that adds to environmental pollution. There are also energy losses associated with high N excretion because urea synthesis from ammonia is an energy costing process. This energy cost may attribute to reducing milk yield, lowering reproductive performance and losses in body condition. Dr Ghulam ***Habib*** *of Pakistan*

We produce enough roughage to feed our cattle, but it is not readily available to all animals. The major challenges are then how to make them available in different seasons and regions. Dr Khan ***Huque*** *of Bangladesh*

Considering the cost of feed and the return from dairying, indigenous cattle and low yielding improved cows producing less than 7 L/day do not return sufficient to be suitable for commercial dairy farming, even with 4 or 5 cows per household. From a practical point of view, the type of dairy animal (and its potential milk yield) must be considered, such as the cost of rearing and the return to the pockets of farmers and in Bangladesh, cows need to be able to produce 10 L/day of milk to be profitable. Dr Quazi ***Huque*** *of Bangladesh*

13.4.3 Technology transfer

Training programs often cover the communication aspect but rarely bring behavioural changes in farmers for adoption. This requires continuous and strong technical backup visits to farmers after training, which unfortunately is frequently missing in many of our extension systems. Once trained, farmers are generally left on their own and seldom visited again. Dr ***Habib*** *of Pakistan*

Farmers are the best economists for their livelihood development. Dr Khan ***Huque*** *of Bangladesh*

Research or technology development should not only be technically sound but also economically beneficial (to producer) and adoptable (by animal owners most being smallholders or resource poor). Dr D. ***Rangnekar*** *of India*

There is a need to develop a system of regular and planned interactions between research and development organisations (both government and non-government), farmer organisations etc. for scientist groups to understand production systems, problems/constraints faced by livestock. Dr D. ***Rangnekar*** *of India*

While economics is an important factor, farmer perception of economics is different from conventional economics, and there are two other equally important factors – convenience and risk perception. So, recommendations may be initially accepted if they are apparently beneficial but if [it] becomes too inconvenient or time consuming, it may not be adopted. Dr D. ***Rangnekar*** *of India*

Women have many other jobs to perform and responsibilities to handle – in a mixed farming, smallholder situation. And in case they smell some risk (such as spoilage during ensiling or treatment), the adoption is difficult since they cannot afford to take further risks. Dr D. ***Rangnekar*** *of India*

Regarding acceptance of improved livestock technology by the village farmers, they will not accept the good words from the scientists, rather they will accept the words from their neighbour fellow farmers who are benefiting by using this technology. So, for proper effective adoption of technology, field and farmer days at the farmer's house are obviously required. Dr S. ***Khan*** *of Bangladesh*

The only way to change behaviour of farmers is through seeing and believing and understanding the direct economic benefits. Dr Quazi ***Huque*** *of Bangladesh*

The model farm concept is beneficial [if] it is established in villages using one of the farmers that are set up for improving their management, making use of the means and ways within their limited resources. There is no point taking resource poor farmers to facilities (such as those established in public or commercial private sectors with expensive state of art technology) that cannot be copied and adopted by them. When farmers see the possibility and benefits of improving farm management under conditions close to their own, they are motivated to adopt. Dr Ghalam ***Habib*** *in Pakistan*

After functioning as a development worker, I have learnt that it is 'a blessing in disguise' for small farmers in that they often resist change thus creating challenges for technical persons. I say this with full knowledge of the way that livestock extension works. Dr D ***Rangnekar*** *of India*

It's possible there is much re-inventing of the wheel going on, even in the vast Bathan dairying lands that feed milk into Milk Vita's Bagharbarighat powder plant; and into other more recently established plants based on the Milk Vita model. Plainly these earlier interventions have not taken root, nor been able to get the right message out to typical milk producers. Perhaps someone might consider collecting the many manuals, guidelines, resource books and the like that dairy interventions (project and research) have churned out over the past half century since independence and distil them into user-friendly, area-specific, best feeding practice for Bangladeshi milk producers – small, medium and large. Dr Brian ***Dugdill*** *of England*

13.5 FAO course on 'Feeding management of dairy cattle in tropical countries'

A new program written by the senior author is being offered by the Animal Production and Health Division of the Food and Agriculture Organization (FAO), in Rome, Italy, as part of their capacity building activities for advisers to tropical SHD farmers. This is a training program in the area of feeding management of dairy stock, both milking cows and young stock. The course has proven very popular as it has been downloaded 457 times and accessed online 226 times over the 12-month period up to December 2015; it has even been translated into French. The course comprises 16 self-learning modules accessible on the internet. Each module has a series of Knowledge Assessment Tasks to assess the learning process.

The objective of the course is for dairy stakeholders to gain a more objective, comprehensive and technical understanding of the complex nature of the feeding management decisions carried out by successful and profitable smallholder and large-scale dairy farmers.

The target audience is:

- extension workers working directly with dairy farmers
- government and agribusiness dairy advisers
- progressive smallholder and/or large-scale dairy farmers/managers
- dairy cooperatives and their support infrastructure educators (both degree and certificate level)
- senior government management and policymakers
- staff of Non-Government Organisations
- teachers and university lecturers.

13.5.1 What does the course contain?

There are 16 modules (lessons). The content covers:

- Principles of dairy nutrition and feeding management as cows move through their annual lactation cycle.
- Quantification of the likely milk responses to changes in intakes of feed nutrients so that a ration can be formulated to achieve a targeted milk yield.
- A practical session on troubleshooting feeding problems.
- Specific aspects of dairy production technology, such as rearing young stock and feeding for improved fertility.

Each module is composed of a set of structured slides. Slides contain learning material, tests and supplementary files for further reading. During the learning process users will be led from one slide to another sequentially, but they can also jump from one to another slide. The contents of each module are generally organised in the following sequence:

1. Welcome.
2. Learning objectives.
3. Introduction.
4. Learning contents.
5. Summary.
6. Self-assessment.
7. Further readings.

Title of each lesson:

1. A brief introduction to tropical smallholder dairy farming.
2. Principles of feeding management.
3. Key nutrients in feeds.
4. Key nutrients in milk.
5. The lactation cycle.
6. Quantifying the nutrient requirements of milking cows.
7. Supplying these key nutrients from forages.
8. Balancing rations with supplements.
9. Formulating a ration.
10. Milk responses to supplements.
11. Troubleshooting feeding problems.
12. Feeding for improved fertility.
13. Feeding pre-weaned calves.
14. Feeding weaned heifers.
15. Measuring body condition score.
16. How feeding management impacts on other aspects of herd management.

13.5.2 How to access the course

The course is available on the FAO website from which it can be easily downloaded onto a computer or it can be accessed online.

The instructions to access it are as follows:

- A video has been developed to present the program. It can be accessed via Google on https://www.youtube.com/watch?v = CcbxNdcgW8c
- Type in http://www.fao.org/elearning/#/elc/en/courses/APH
- Select the course, which is called 'Feeding management of dairy cattle in tropical countries'.
- First-time users must register with a User name, Password and some personal details such as your organisation, email address, gender, country of origin and identifying the hidden letters.
- Once you have successfully registered and re-entered your User name and Password, you have the option of either running the course online or downloading it onto a folder in your computer.
- To download it, you need to select a specific folder, then once the 21.7 MB has been downloaded, you choose 'Extract All' from the options and select a destination folder from which to actually run the course.
- You then find and activate the file 'start-course.exe' which takes you to the opening page on which you click on 'Click here to start now', then follow the simple instructions to open up any of the 16 modules.

13.6 Future directions for SHD production in the humid tropics

The following are some of the key issues in feeding management on SHD farms that require further attention at both the research and extension levels. The senior author wrote out these issues over 10 years ago, as a 'wish list' for his first book (Moran 2005). Since they are still highly relevant to Asia's developing dairy industries, we believe that it is worth listing them again.

13.6.1 Feed nutrients

- As forage supplies are of paramount importance, inexpensive, year-round and sustainable supplies of quality roughages must be developed for every system of SHD farming.
- Because dry season forages are poorer in nutritive value, conserved excess wet season forages often form the basis of most profitable dairy systems.
- Tropical forages are high in fibre, so farmers will always require concentrate supplements.
- Energy will continue to be the major limiting feed nutrient in smallholder dairying. Improving the energy status of milking cows will be of benefit to both the production of milk and milk solids.
- Home-grown forages will almost always be cheaper sources of feed nutrients than purchased forages, particularly when managed to provide optimum yields and quality.

13.6.2 Feeding management

- The high nutrient demands of milking cows negate many of the recommended feeding systems based on chemically treated low quality roughages.
- Agro-industrial by-products will always form the basis of concentrate supplements, because of increasing demand for land to grow crops for human consumption.
- With continuing emphasis on increasing the domestic milk production in all SE Asian countries, it is important not to 'overstock' any developing dairy region. Feed audits, particularly of 'home-grown' forage supplies, should be undertaken and adhered to when projecting optimum numbers of dairy stock for any particular region.
- In addition to encouraging farmers to improve agronomic practices on their own smallholder farms, 'home-grown' forages can be produced on communal areas, utilising the expertise of cooperative staff to oversee management to optimise fertilising, harvesting and if necessary, conservation of quality forages for use by nearby farmers. If there are economic justifications for cooperatives to bulk purchase ingredients and formulate concentrates, similar benefits could arise with managing forage production areas.

- High yielding cows are very susceptible to heat stress. Every effort should be made to alleviate the adverse effects of heat stress on appetite and fertility, hence milk production and profitability.
- An underutilised measure of feeding management is the persistency of the lactation curve. It is one thing to feed for high peak yields, but it is just as important for good milk yields to be maintained throughout the lactation. Farmers should feed cows to allow milk yields to fall by no more than 10% decline from peak yield per month.
- The simplest 'take home message' from all the technical books, seminars, demonstration farms and farmer discussion groups relevant to tropical dairy farming is **feed fewer cows better**.

13.6.3 Other issues

- There is little sense in importing 'improved genetics' or growing out bigger heifers if they cannot be fed well when milking.
- Dairying is a business, so feeding decisions should be based on logical and appropriate economic information with 'Milk income above feed costs' being the single most useful measure of success. Although this measure does not take into account non-feed costs, a farm with high costs of production usually has high non-feed as well as high feed costs.
- Improving the knowledge of basic ruminant nutrition will greatly assist many dairy advisers to formulate more profitable milking rations, because generic recipes are notoriously unreliable.
- Farm development is often limited by inefficient (and even inappropriate) technology transfer in that extension procedures do not always acknowledge farmer skills and adult learning principles.

Concern is often expressed that most SE Asian countries are not in a position to develop large-scale intensive livestock industries without the importation of feedstuffs. This certainly applies to pig and poultry but for dairying, another relatively intensive industry, the future may not be for large farms requiring high tonnages of fresh forages and imported concentrates. Unlike Europe, North America and Australasia, where dairying evolved from smallholders to larger and often corporate farms, we believe the future for dairying in SE Asia is still with the smallholder sector. For these systems, fresh quality forages can often be sourced close by while local by-products can form the bulk of the concentrate formulations.

Even in one of the most densely populated islands in the world, Java, there are large tracts of underutilised land such as forest plantations and aging coffee and rubber plantations, where SHD farmers currently can freely harvest limited amounts of forage for their stock. In recognising their potential, Indonesian policymakers place high priority in developing such resources through providing greater access to farmers and plantings of improved forage species (Burrell and Moran 2004). With better feeding management skills, smallholder farmers can profitably increase raw milk supplies, to the nutritional benefits of their fellow consumers and the improved food security of their nations.

13.7 In summary

In spite of several decades of dairy farming in the tropical developing countries of SE Asia, the productivity of SHD farming has remained relatively low due in part to the lack of appropriate dairy research and application. Small farmers, due to their socio-economic and agro-economic conditions being greatly different to those in developed countries, cannot readily adopt the science and technology available in developed countries. Even the most appropriate technology is rarely transferred to smallholders *en masse* due to a general lack of effective extension support services. There needs to be large-scale institutional support to facilitate dairy industry growth through mechanisms such as providers of farmer credit, farmer training centres, well-equipped milk collection centres, processing and marketing facilities, farmer cooperatives or groups and appropriate research and extension infrastructures and methodologies. School milk programs have been successful in encouraging the development of SHD farming by promoting milk drinking to improve health among children, particularly in rural areas. It is then essential for any production technology being transferred to these farmers to be relevant to their needs as well as being economically and practically feasible, given their local support networks of dairy cooperatives, advisers (government and agribusiness), creditors and milk handling and processing infrastructures. Production technology skills should automatically lead onto animal welfare skills.

Despite the complexities of establishing and successfully operating large-scale, intensive dairy feedlots, this has been and is being undertaken in developing countries such as China and Vietnam. Granted the advice and assistance from intensive dairy specialists from more developed Western dairy industries have been sought and utilised, however, these successful models demonstrate that even in the humid tropics, such production systems have a viable, hence long-term future. Many other SE Asian countries are also following down this line with fewer, though equally successful, dairy feedlots, such as Thailand, Indonesia, India, Pakistan and Malaysia. The more consistent production of high quality raw milk arising from these more controlled dairy production systems provides the basis of satisfying consumer demands for both liquid milk and the more expensive, value-added dairy products. Such large-scale dairy systems will increase in both numbers and product supply, being driven by national policies for dairy self-sufficiency and increasing affluence of local populations.

Good dairy cattle husbandry includes the provision of appropriate resources of feed and shelter, effective management and sympathetic stockmanship. These include:

- Physical resources necessary to ensure proper feeding, housing and hygiene
  - well-constructed, properly replenished feed stores
  - accommodation that is hygienic, physically and thermally comfortable and unlikely to cause injury
  - facilities for routine preventative medicine and the care of individual sick animals
- Strategic management designed to address the physiological, health and behavioural needs of the animals
  - feeding, production, health and welfare plans devised and implemented with professional advice as appropriate to the needs of the system and the individual animals
  - comprehensive records relating to feeding, production, health and welfare
- Competent stockmanship, sympathetic to the day-to-day needs of the stock
  - a skilled empathetic approach to animal handling
  - early recognition and attention to any signs of disease or injury
  - work practices that encourage competent and caring stock keepers and which give them the time to develop empathy with the animals in their care.

A competent, welfare-minded dairy farmer is then one who extends his farming skills to cover cow psychology as well as cow production technology and farm business management. In essence, he should be able to put himself 'inside the cow's skin' to develop the ability to 'think like a cow'. Optimum cow comfort is just as important as the provision of the most appropriate resources on-farm, because if the cow is stressed, their appetite for feed nutrients can be as much a limiting factor as their nutrient supplies.

Appendix 1. Temperature Humidity Index

The following table presents the Temperature Humidity Index, calculated from temperature (in °Fahrenheit or Centigrade) and relative humidity (%), highlighting its potential effects on cow heat stress and therefore, on the cow's performance.

Temperature		Relative humidity																				
Deg F	Deg C	0	5	10	15	20	25	30	35	40	45	50	55	60	65	70	75	80	85	90	95	100
72	22.2																				72	72
73	22.8																		72	72	73	73
74	23.3																72	72	73	73	74	74
75	23.9														72	72	73	73	74	74	75	75
76	24.4												72	72	73	73	74	74	75	75	76	76
77	25.0											72	72	73	73	74	74	75	75	76	76	77
78	25.6										72	73	73	74	74	75	75	76	76	77	77	77
79	26.1									72	73	73	74	74	75	76	76	77	77	78	78	79
80	26.7							72	72	73	73	74	74	75	76	76	77	78	78	79	79	80
81	27.2						72	72	73	73	74	74	75	76	77	77	78	78	79	80	80	81
82	27.8						72	73	73	74	75	75	76	77	77	78	79	79	80	81	81	82
83	28.3					72	73	73	74	74	75	76	77	78	78	79	80	80	81	82	82	83
84	28.9				72	73	73	74	75	75	76	77	78	78	79	80	80	81	82	83	83	84
85	29.4			72	72	73	74	75	75	76	77	78	79	79	80	81	81	82	83	84	84	85
86	30.0			72	73	74	74	75	76	77	78	78	79	80	81	81	82	83	84	84	85	86
87	30.6		72	73	73	74	75	76	77	77	78	79	80	81	81	82	83	86	85	85	86	87
88	31.1	72	72	73	74	75	75	76	77	78	79	80	81	81	82	83	84	85	86	86	87	88
89	31.7	72	73	74	75	76	76	77	78	79	80	81	82	83	84	85	86	86	87	88	89	89
90	32.2	72	73	74	75	76	77	78	79	79	80	81	82	83	84	85	86	86	87	88	89	90
91	32.8	73	74	75	76	76	77	78	79	80	81	82	83	84	85	86	86	87	88	89	90	91
92	33.3	73	74	75	76′	77	78	79	80	81	82	83	84	85	85	86	87	88	89	90	91	92
93	33.9	74	75	76	77	78	79	80	80	81	82	83	84	85	86	87	88	89	90	91	92	93
94	34.4	74	75	76	77	78	79	80	81	82	83	84	85	86	87	88	89	90	91	92	93	94
95	35.0	75	76	77	78	79	80	81	82	83	84	85	86	87	88	89	90	91	92	93	94	95
96	35.6	75	76	77	78	79	80	81	82	83	85	86	87	88	89	90	91	92	93	94	95	96
97	36.1	76	76	78	79	80	81	82	83	84	85	86	87	88	89	91	92	93	94	95	96	97
98	36.7	76	77	78	79	80	82	83	84	85	86	87	88	89	90	91	93	94	95	96	97	98
99	37.2	76	77	79	80	81	82	83	84	85	87	88	89	90	91	92	93	94	96	97	98	99
100	37.3	77	78	79	80	82	83	84	85	86	87	88	90	91	92	93	94	95	97	98	99	
101	38.3	77	79	80	81	82	83	86	86	87	88	89	90	92	93	96	95	96	97	99		
102	38.9	78	79	80	81	83	86	85	86	87	89	90	91	92	96	95	96	97	96			
103	39.6	78	79	81	82	83	86	86	87	88	89	91	92	94	95	96	97					
104	40.0	79	80	81	82	86	85	86	88	88	90	91	93	94	95	96						
105	40.6	79	80	82	83	86	86	87	88	89	91	92	93	96	96	97						
106	41.1	80	81	82	86	85	86	88	89	90	91	93	94	95	97	98						
107	41.7	80	81	83	86	85	87	88	89	91	92	94	95	96	98							
108	42.2	81	82	83	85	86	87	89	90	92	93	94	96	97								
109	42.3	81	82	86	85	87	88	89	91	92	94	95	96	98								
110	43.3	81	83	86	86	87	88	90	91	92	94	96	97									
111	43.9	82	83	85	86	88	89	91	93	94	95	96	98									
112	44.4	82	86	85	87	88	90	91	94	94	96	97										
113	45.0	83	86	86	87	89	90	92	95	95	96	96										
114	45.4	83	85	86	88	89	91	92	94	96	97											
115	46.1	86	85	87	88	90	91	94	95	96	98											
116	46.7	86	86	87	89	90	92	94	95	97												
117	47.2	85	86	88	89	91	93	94	96	98												
118	47.3	85	87	88	90	92	93	95	97													
119	48.3	85	87	89	90	92	94	96	97													
120	48.9	86	88	89	91	93	94	96	98													
121	49.4	86	88	90	92	93	95	97														

Five comfort zones for milking cows

- A <72 No stress
- B 72–78 Mild stress
- C 78–89 Severe stress
- D 89–98 Very severe stress
- E >98 Dead cows

Appendix 2. Signals for assessing healthy and sick cows

Signals for healthy cows

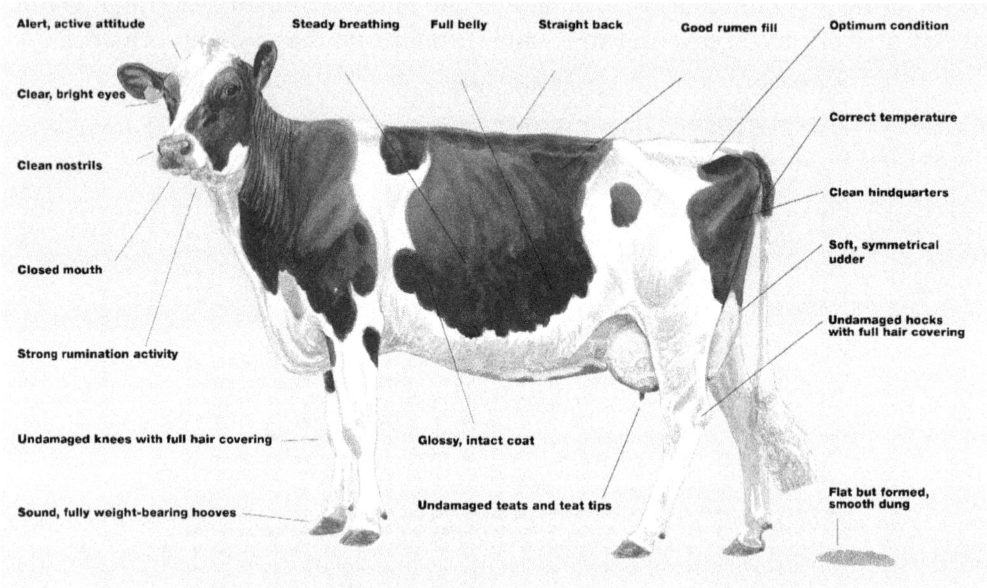

Cow signals to assess health and welfare

Cow's anatomy	Observation and interpretation
Whole body	**Alert, active attitude** Distracted attitude: indicates poor health, low energy status and possibly rumen acidosis
	Steady breathing Rapid and shallow: heat stress or pain. Sometimes at start of rumination period as well Normal: 10 to 30 times per minute
	Optimum condition Too thin: inadequate energy intake Too fat: excessive energy intake Normal: good flesh cover with a little fat
	Glossy intact coat Dull coat: poor health or nutrition Skin injuries: cause and result of agitation and reduced disease resistance
	Correct body temperature Too high (>39.0°C): fever Too low (<38.0°C): milk fever or serious illness Normal (between 38.0 and 38.5°C): healthy
Head	**Clear, bright eyes** Distracted attitude indicates poor health, low energy status and possibly rumen acidosis
	Clean nostrils Mucus with pus/blood and skin injuries: skin of the nose inflamed indicates virus or cold Clear mucus does not indicate much
	Closed mouth Some drooling: usually hunger Lot of drooling: swallowing problems or mouth pain Coughing: due to cold air, dust or disease
	Strong rumination activity Reduced chewing: diet lacks effective fibre Spitting out the cud: tooth problems, prickly bits in feed Normal: 55 to 75 chews per cud
Forequarters	**Undamaged knees with full hair covering** Bare knees: scraping on ground while getting up Swollen knees: bruising when getting up, lack of space in stall
	Sound, fully weight-bearing hooves Tiptoeing, standing on tips of hooves Injured or swollen coronary band Eczema or scabs in interdigital space
Abdomen	**Full belly** Belly too empty: hadn't eaten enough last week Take account of the size of the calf, if any
	Good rumen fill Too empty: hasn't eaten enough yesterday No discernible layered structure (apple shaped): not enough fibre in diet
	Straight back Arched back: painful hooves or physical wear and tear Injuries: usually bruising against stall partition

(Continued)

Cow signals to assess health and welfare (Continued)

Cow's anatomy	Observation and interpretation
Hindquarters	**Clean hindquarters** Dung on both sides of rump: dung too thin Asymmetric soiling: environment too dirty
	Soft, systematic udder Hard: due to oedema around calving or mastitis (painful) Enlarged quarter: active mastitis Shrunken quarter: previous mastitis
	Undamaged teats and teat tips Trodden teats: too much agitation, stalls too narrow or too slippery Check milking machine and technique if you see calloused teat tips, swelling, redness or tiny blood spots
	Undamaged hocks with full hair covering Bare hocks: scraping on stall floor, lack of grip Thick hocks: lack of stall space, stall floor too hard Scabs: inflammation due to dirt or moisture
	Flat but formed, smooth dung Long stems: insufficient rumination activity Not too loose or too firm: always relate to ration components (e.g. grazed pastures) and lactation stage (dry v peak lactation) Use dung feedback to assess the diet: feeding methods, feed intake, digestion, water intake and health

Signals for sick and distressed cows

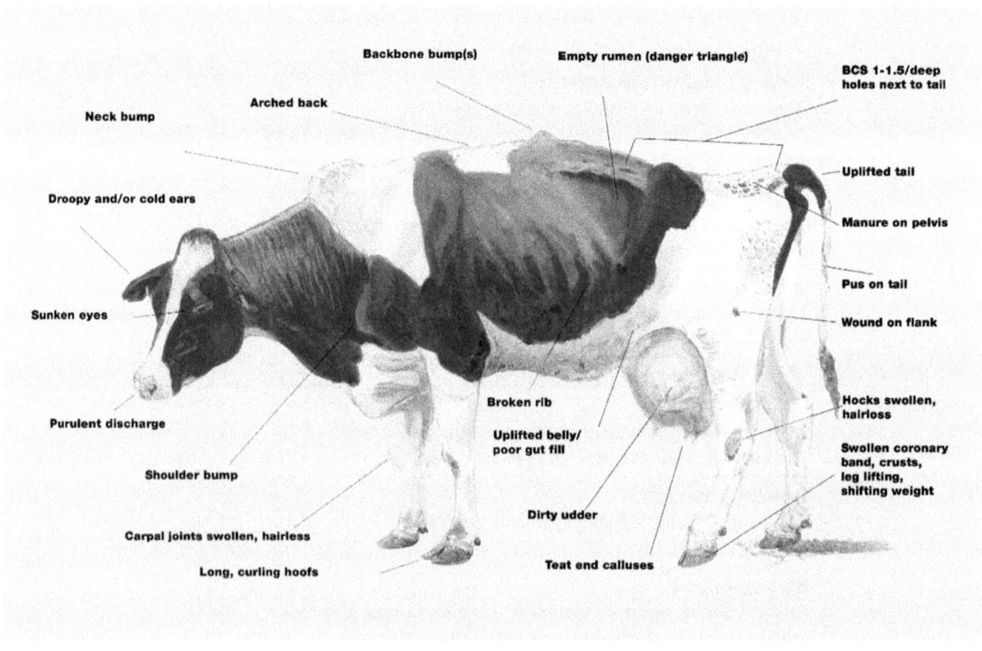

Sickness and distress check list

Cow's anatomy	Observation and interpretation
Head	Sunken eyes: sick
	Droopy and/or cold ears: sick
	Purulent nasal discharge: rhinitis or chronic lung problem
Fore quarters	Neck bump: neck rail feed fence too low
	Shoulder bump: poorly designed feed fence/feed too far away
	Long curling hoofs: hoof trimming too late or not at all
	Carpal joints swollen, hairless: stall surface too hard, too little head space
Abdomen	Arched back: lame
	Backbone bumps: lying against stall divider
	Empty rumen: has eaten too little
	Broken rib: stall divider
	Uplifted belly/poor gut fill: pain/poor feed intake for days
Hind quarters	Body condition score of 1 to 1.5 and deep holes next to tail: long-term physical problems
	Dirty udder: dirty resting area and/or floors
	Teat end callouses: milking machine problems
	Uplifted tail: pain in birth canal
	Manure on pelvis: diarrhoea
	Pus on tail: endometritis
	Wound on flank: wet resting surfaces
	Swollen hocks and hair loss: stall surfaces too hard, abrasive or not enough grip
	Swollen coronary band, crusts, leg lifting, shifting weight: hoof diseases

Appendix 3. Questionnaire for Indonesian dairy farmers

Farm Name:
Owner Name & Age:
Interview Date:

Location:
Contact Details:
Farm Code:

1. GENERAL QUESTIONS

Code	Question	Answer
1a	What % of your farm income comes from dairy farming?	
1b	What other farming enterprises do you undertake?	
1c	What is the total area of your dairy enterprise (in ha)?	
1d	What is the total dairying area that you own, or are paying off (ha)?	
1e	How much area are you leasing (ha)?	
1f	What proportion of your farm do you use to produce fodder for your dairy enterprise?	
1g	How many years have you been dairy farming?	
1h	How many years have you been dairy farming on this particular farm?	
1i	When was the last time you constructed improvements to your dairy shed and other farm facilities?	
1j	How much of your forage production land do you irrigate?	
1k	How big is your dairy shed?	
1l	What is the distance to your nearest milk collection centre?	
1m	Are you a member of your local farmers' group (dairy cooperative, milk supplier group)?	
	Plans for the future	
1n	Do you intend to produce milk next year?	
1o	If so, do you plan to produce more milk than this year?	

(Continued)

1. GENERAL QUESTIONS (Continued)

Code	Question	Answer
1p	Do you expect a change in unit milk returns next year? Answer: 1, no change; 2, increase; 3, decrease	
1q	Do you expect a change in concentrate unit prices next year? 1, no change; 2, increase; 3, decrease	
1r	Next year, do you want to have more milking cows on your current forage production area?	
1s	Will you be dairy farming in 5 years' time?	
1t	Do your children want to follow you on the farm?	
1u	Is dairy farming expanding (1), static (2) or declining (3) in your area?	

2. FARM ASSETS & LIVESTOCK

Number and value of dairy livestock

	Number of dairy livestock	Value (Rp)	Number now	Number 12 months ago
2a	Milking cows (adult)			
2b	Dry cows (pregnant)			
2c	Dry cows NON pregnant			
2d	Heifers (weaning to point of calving)			
2e	Heifer calves			
2f	Bull calves			
2g	Other dairy stock (Bulls)			

Value of farm assets

Code	Question	Answer
2f	Value of buildings	
2g	Value of land owned (/ha)	
2h	Farm vehicle (and % used for farm business)	
2i	Effluent pump	
2j	Heat stress system (sprinklers, fans)	
2k	Motorbike	
2l	Trailer	

(Continued)

Value of farm assets (Continued)

Code	Question	Answer
2m	Tractor	
2n	Milking machines	
2o	Rubber mats	
2p	Milk cans and milk vat(s)	
2q	Grass chopper	
2r	Feed mixer (TMR)	
2s	Other dairy-related farm assets	
2t	Generator	
2u	Other	
2v	Other	

Farm loans

Code	Loan number	Initial value	Years to run	Annual interest	Annual repayment
2w					
2x					
2y					

3. FARM INCOME

Annual income (or estimated monthly income × 12)

Code	Question	Answer
	Milk sales	
3a	Yesterday's milk production (kg for the farm)	
3b	Average milk price (Rp/kg)	
3c	Annual milk return (Rp)	
	Livestock sales	
3d	Number of stock sold in last 12 months	
3e	Average value (Rp)	
3f	Annual return from stock sales (Rp)	
	Other dairy farm sales over last 12 months	
3g	Manure	
3h	Forages	
3i	Planting material	
3j	Other dairy-related sales	

4. VARIABLE COSTS: Herd and shed costs

Code	Question	Answer
	Artificial insemination	
4a	Veterinarian visits	
4b	Drugs and other costs	
	Rearing young stock (calves and yearlings)	
4c	Fresh milk	
4d	Calf milk replacer	
4e	Concentrates	
4f	Forage	
	Health care	
4g	Vaccinations	
4h	Other drugs	
4i	Veterinarian visits	
4j	Other health care costs	
	Milk harvesting	
4k	Rubber ware	
4l	Detergents and sanitisers	
4m	Repairs and maintenance of milking equipment	
4n	Transportation of milk to collection centre	
4o	Towels	
4p	Other milk harvesting costs	

5. VARIABLE COSTS: Feed costs

Code	Question	Answer
	Concentrates	
5a	Factory concentrates (pellets or loose mix)	
5b	Protein meals	
	Other feed supplements	
5c	Mineral blocks	
5d	Mineral and vitamin premixes	
5e	Other feed supplements	
	By-products	
5f	Brewer's grain	
5g	Molasses	
5h	Other by-products (PKE)	
	Purchased forages	
5i	Rice straw	
5j	Native grass	
5k	Corn stover	
5l	Other forages	
	Home-grown forages	
	1. High quality forages	
5m	Land preparation	
5n	Seed planting material	
5o	Urea fertiliser	
5p	NPK fertiliser	
5q	Organic manure	
5r	Petrol for irrigation scheme water purchased	
	Processing and storage of forages	
5s	Urea treated rice straw	
5t	Making silage	
5u	Harvesting crops	

(Continued)

5. VARIABLE COSTS: Feed costs (Continued)

Code	Question	Answer
	2. Crops grown for dairy enterprise	
	Crop production costs	
5v	Land preparation	
5w	Seed planting material	
5x	Urea fertiliser	
5y	NPK fertiliser	
5z	Organic manure	
5aa	Petrol for irrigation	
	Processing and storage of forages	
5bb	Urea treatment	
5cc	Making silage	
5dd	Harvesting crops	
	Feed-related machine costs	
5ee	Petrol and oil for forage transport	
5ff	Electricity for forage chopper	
5gg	Repairs and maintenance of forage machinery	
	Non-feed or milk transport-related machine costs	
5hh	Other petrol and oil	
5ii	Other electricity	

6. OVERHEAD COSTS

Code	Question	Answer
6a	How many labour units do you employ?	
6b	What is their monthly salary (/mth)?	
6c	How many family members work on the dairy farm?	
6d	On average, how many hours do they work each week?	
	1. Cash overhead costs	
6e	Paid labour	

(Continued)

6. OVERHEAD COSTS (Continued)

Code	Question	Answer
6f	Finance cost on interest of loans	
6g	Tax on land owned by farmer	
6h	Land Rental	
6i	Payments to cooperative (Membership etc. if applicable)	
6j	Office administration	
6k	Miscellaneous cash overhead costs	
	2. Imputed overhead costs	
	Family labour	
6l	Basic payment for family labour	
6m	Management	
6n	Administration	
	Depreciation of farm assets	
6o	Shed	
6p	Biogas plant	
6q	Milking machine	
6r	Forage chopper	
6s	Effluent pump	
6t	Irrigation system (incl. pump)	
6u	Motorbike	
6v	Trailer	
6w	Small tractor	
6x	Heat stress system	
6y	Farm vehicle	
6z	Milk cans	
6aa	Grass chopper	
6bb	Feed mixer	
	Capital purchases	
6cc	Loan repayments	
6dd	Capital improvements	
6ee	Capital purchases	

7. OTHER FARM MANAGEMENT QUESTIONS

Code	Question	Answer
	During this visit how many stock do you have:	
7a	Mature/heifer milking cows, currently milking	
7b	Mature/heifer milking cows, currently dry	
7c	Growing weaned heifers	
7d	Milk-fed calves	
7e	Bulls (growing and mature)	
7f	Other dairy or beef stock	
7g	Proportion of forage sourced off-farm	
7h	Do you use fresh milk (1) and/or calf milk replacer (2) for rearing the calves?	
	For fertilisers do you use:	
7i	Shed effluent	
7j	Inorganic fertilisers	
	Types of concentrates fed to milking cows:	
7k	Factory formulated concentrates	
7l	Home mixed concentrates	
	Farm income generation	
7m	Number of cows in milk	
7n	Rolling herd average milk yield	
7o	Average per cow production	
7p	Typical peak milk yield	
7q	Best peak milk yield	
7r	Rate of milk decline	
7s	Milk fat content	
7t	Milk protein (or SNF) content	
7u	Total milk solids content	
7v	Milk quality (TPC)	
7w	Typical milk price for good quality milk	
7x	Milk price for farm output over last 7 days	

(Continued)

7. OTHER FARM MANAGEMENT QUESTIONS (Continued)

Code	Question	Answer
7y	Typical calving interval (months)	
7z	Average number or DIM between calving	
7aa	Average number of dry days before calving	
7bb	Submission rate (number of days until first mating)	
7cc	Number of days post-calving to achieve conception (100 DIC)	
7dd	Typical conception rate for 1st insemination	
7ee	If using AI, number of services per conception	
7ff	Overall conception rate	
7gg	Typical age at 1st calving (mth)	
7hh	Typical losses of cows and calves before birth. Abortion rate.	
7ii	Typical morbidity rates (sickness) before weaning	
7kk	Typical mortality rates of calves before weaning	
7ll	Typical mortality rates post-weaning	
7mm	How long do you feed milk to the calves (weeks)?	

8. FEEDS SUPPLIED TO DAIRY HERD

List the **daily** feed types and quantities for the various animal groups

Lactating cows	Feed	Quantity (kg)	Rp/kg	Dry matter (%)	Metabolisable energy (MJ/kg DM)	Crude protein (%)

FEEDS SUPPLIED TO DAIRY HERD (Continued)

Dry cows	Feed	Quantity (kg)	Rp/kg	Dry matter (%)	Metabolisable energy (MJ/kg DM)	Crude protein (%)

Yearlings	Feed	Quantity (kg)	Rp/kg	Dry matter (%)	Metabolisable energy (MJ/kg DM)	Crude protein (%)

Calves	Feed	Quantity (kg)	Rp/kg	Dry matter (%)	Metabolisable energy (MJ/kg DM)	Crude protein (%)

9. OTHER QUESTIONS ABOUT GENERAL FARM MANAGEMENT

Is mastitis a problem and do you use Dry Cow Therapy at drying off?	
How many cows have mastitis today?	
Do you practise a routine vaccination program?	
Do you have a worm treatment program?	
Is there a special calving down area?	
Is there a water supply scheme on the farm?	
What % of cows are ruminating right now?	
Do the cows appear comfortable?	
Is lameness a problem?	
What is the farmer's age?	
Do you use a California Mastitis Test kit and how often?	
Does the farm maintain herd health and reproduction records?	
Does the farm maintain financial records?	

	What are the three biggest problems in the management of your dairy farm?
1	
2	
3	

Appendix 4. A floor plan of a 50-cow dairy shed in Pakistan

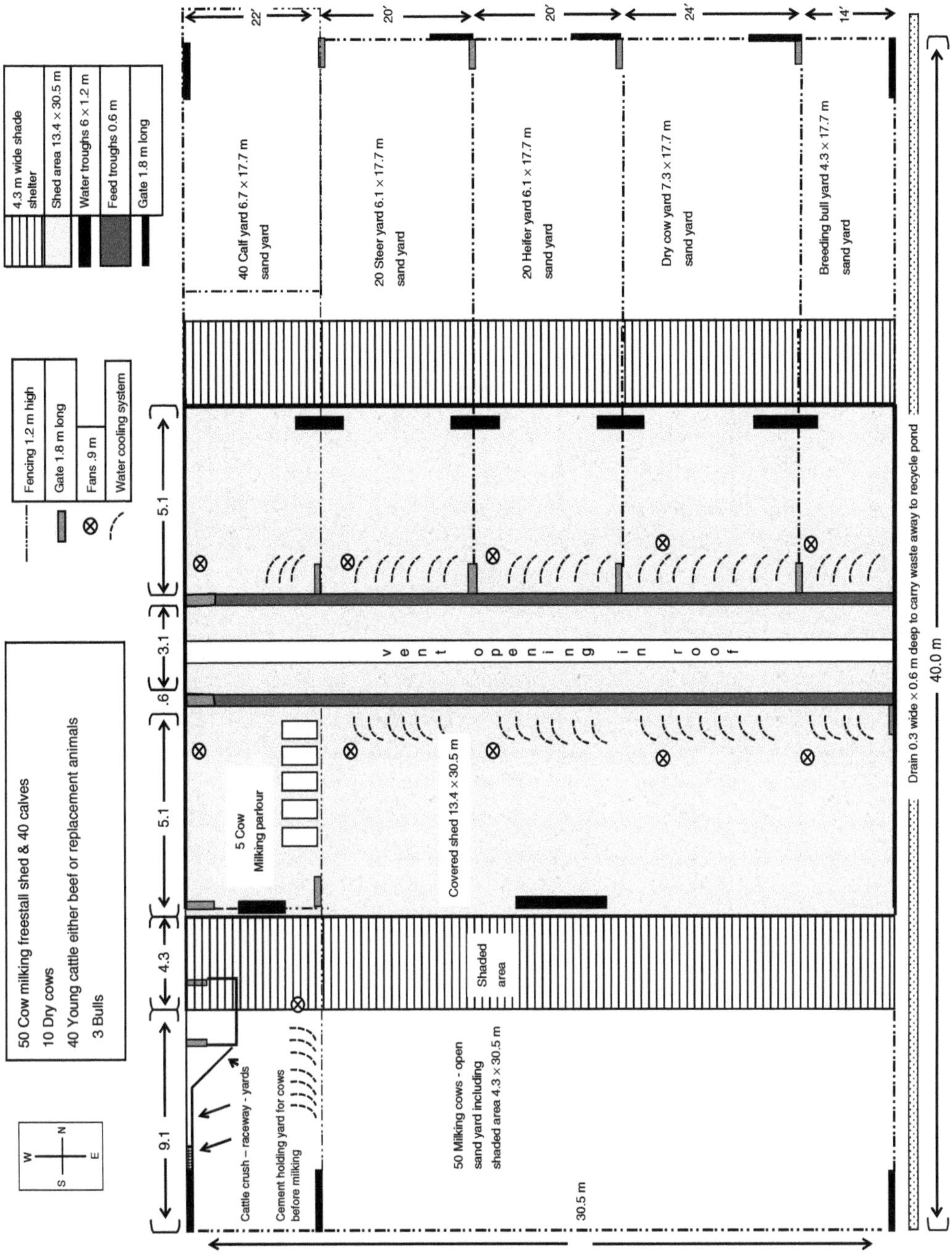

Appendix 5. A plan for a free stall shed

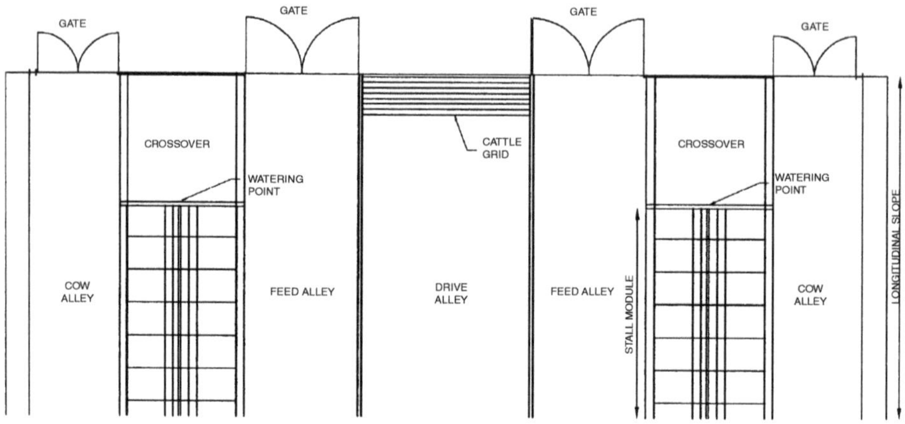

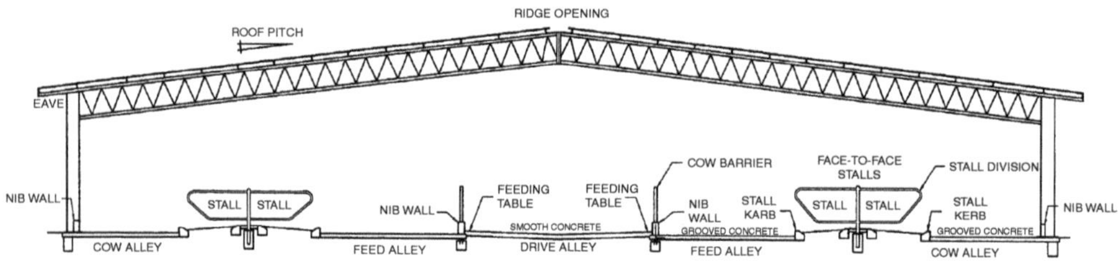

Appendix 6. A plan for cooling a free stall shed

This shed is not a tunnel ventilation barn (see Chapter 5).

Note that these linear measurements are in feet, rather than metres as this is a US design (1 foot = 0.305 m).

Therefore:

- The continuously open ridge vent along the roof top is 76 cm wide.
- The roof pitch is 4%.
- The open sides of the shed are 3.7 m high.
- The shed is 30.5 m wide.
- The concrete feed passage is 5.5 m wide.
- The double free stalls on each side of the shed are 4.6 m wide.
- The cow alleys are 4.3 and 3.7 m wide.
- The fans over the feed alley are 91 cm in diameter, spaced at 9.1 m.

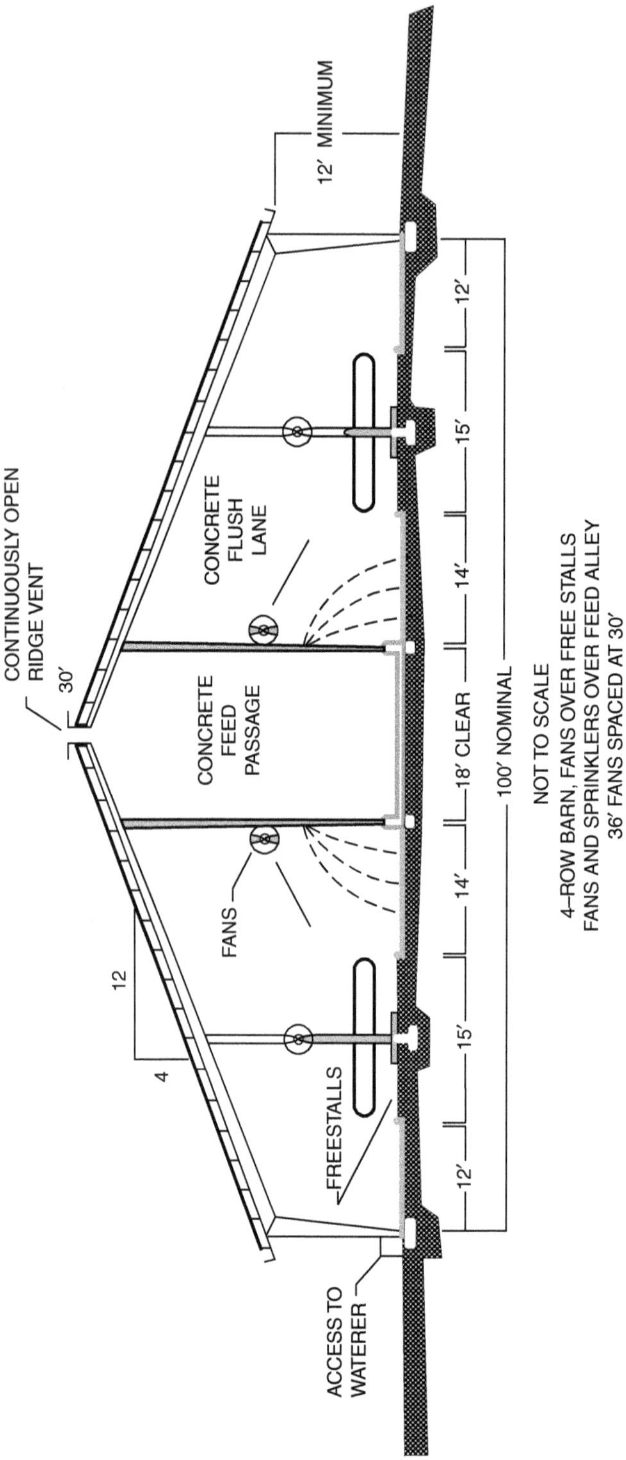
CONTINUOUSLY OPEN
RIDGE VENT
30′
12′ MINIMUM
12
4
CONCRETE
FLUSH
LANE
CONCRETE
FEED
PASSAGE
FANS
FREESTALLS
ACCESS TO
WATERER
12′
15′
14′
18′ CLEAR
14′
15′
12′
100′ NOMINAL
NOT TO SCALE
4–ROW BARN, FANS OVER FREE STALLS
FANS AND SPRINKLERS OVER FEED ALLEY
36′ FANS SPACED AT 30′

Abbreviations

Common abbreviations

Other less common abbreviations are included in the glossary.

<	less than
>	greater than
$	dollar
A	Australian
AI	artificial insemination
C	Centigrade
c	cents
Ca	calcium
cm	centimetre
d	day
F	Fahrenheit
ft	foot
g	gram
ha	hectare
hd	head
hr	hour
Hz	hertz (a measure of sound frequency)
J	joule
K	kilo or thousands
kg	kilogram
L	litre
lb	pound
lux	lux (a measure of light intensity)
LWT	live weight
M	mega or millions
m	metre
MCal	megacalories
mg	milligram
min	minute
MJ	megajoule
ml	millilitre
mm	millimetre
MT	megatonnes
mth	month
N	nitrogen
P	phosphorus
pa	per annum
ppm	parts per million
SE	South-East (Asia)
T	tonnes
US	United States of America
yr	year

Units of measurements

1. Conversion of Imperial units to metric units

Length:

1 inch = 25.4 mm
1 foot = 30.5 cm
1 yard = 0.91 m
1 mile = 1.61 km

Volume:

1 cu ft = 0.028 cu m
1 pint = 0.57 L
1 gallon = 4.54 L
1 bushell = 36.4 L
1 acre foot = 1.23 Ml (megalitre)

Area:

1 acre = 0.40 ha
1 sq mile = 2.59 sq km

Weight:

1 ounce = 28.3 g
1 pound = 0.454 kg
1 hundred weight = 50.8 kg
1 long ton = 1017 kg (2240 lb)

Energy:

1 calorie = 4.19 J

Density:

1 lb/ft^3 = 0.063 kg/m^3

Rate:

1 gallon/acre = 11.23 l/ha
1 pound/acre = 1.12 kg/ha
1 gallon/ton = 4.17 l/tonne

Pressure:

1 pound/sq in (psi) = 1.45 kPa (kilopascals)

Yield:

1 lb/ac = 1.12 kg/ha

Temperature:

1°F = ((9/5) * C) + 32 – conversion
1°F = 0.56°C
50°F = 10.0°C
60°F = 15.6°C
70°F = 21.1°C
80°F = 26.7°C
90°F = 32.2°C
100°F = 37.8°C
110°F = 43.3°C

2. Conversion of US units to metric units

Volume:

1 gallon = 3.79 L
1 bushell = 35.2 L

Weight:

1 hundredweight = 45.4 kg
1 short ton = 907 kg (2000 lb)

Milk prices:

$10/hundredweight = 22.0 c/L

Forage maize yields @ 30% DM:

25 ton fresh weight/acre = 16.8 t DM/ha

Food energy:

1% unit TDN = 0.185 MJ/kg DM of metabolisable energy
30% TDN = 3.7 MJ/kg DM of ME
40% TDN = 5.5 MJ/kg DM of ME
50% TDN = 6.4 MJ/kg DM of ME
60% TDN = 7.4 MJ/kg DM of ME
70% TDN = 8.3 MJ/kg DM of ME
80% TDN = 9.2 MJ/kg DM of ME
1 MCal/lb = 9.22 MJ/kg
1 MCal/kg = 4.19 MJ/kg

Glossary

Terms in this glossary are defined in the context of their use in this book.

acidosis (or grain poisoning) This occurs when rumen pH falls too low through overproduction of lactic acid (the end product of grain digestion) which reduces feed digestion and sometimes causes death.

ad lib* or *ad libitum Fed to appetite.

age at first calving (AFC) A good indicator of heifer management in year-round calving herds. In seasonal-calving herds it is usually predetermined at about 24 months of age.

Animal Production and Health Commission for Asia and the Pacific (APHCA) The FAO's livestock program for Asia and the Pacific.

appropriate technology A term developed to describe practical solutions to problems that could be readily accepted, hence undertaken by farmers – particularly traditional farmers with minimal resources.

asset Anything controlled, owned or not, fixed or current, tangible or intangible that is of value to its possessor.

balance sheets Record of assets and liabilities to determine net worth of the business. Usually undertaken at beginning of each financial year to assess growth in net worth, or wealth creation.

Best Management Practice (BMP) A description of the most suitable procedures for undertaking a set of tasks to develop a checklist for planning various activities on the farm. Basically, it is 'saying what you do, doing what you say, then recording what you have done'.

biosecurity A strategy of management practices to prevent introduction of disease and pathogens to the operation and to control spread within the operation.

body condition A subjective estimate of the amount of subcutaneous fat between the pin bones and the tail head, over the hip and covering the lumbar vertebrae.

budget A detailed statement of a future plan of action detailing the expected costs and benefits.

business The money-making activities of a single enterprise farm or a collection of enterprises on a mixed farm.

calf milk replacer (CMR) A powder constituted to substitute for whole milk when fed to young calves.

California Mastitis Test (CMT) The name given to the test for subclinical mastitis taken on freshly harvested milk.

calving interval The average time period between consecutive calvings in a dairy herd. The target is 12 months, although this is rarely achieved.

calving to conception interval The time period between when a cow has a calf and she next becomes pregnant. It is also called days open.

capital Items that contribute to production over at least a medium-term period, such as 12 months. Examples are tractor, land, fertiliser, farm structures and equipment. Capital investment is using funds to acquire capital assets.

cash An exact known amount you pay or is paid to you, in the form of currency, a cheque, transfer or credit card. It is not imputed, nor includes depreciation of assets.

cash flow statement A statement of cash inflows less cash outflows to indicate cash surplus or deficit.

cash operating surplus (COS) A measure of cash profit which quantifies the sum of all cash flows on the farm.

colony forming units (CFU) The measure of the level of bacterial contamination in raw milk when analysed for Total Plate Count (TPC).

colostrum The milk produced by cows for the first two milkings post-calving which contains high levels of nutrients and immunoglobulins for transferring immunity onto newborn calves.

comfort zone The range of air temperature when there is no measurable fluctuation in physiological processes of cattle.

conception rate The proportion of the total number of services or inseminations that result in pregnancy.

contract heifer rearing The term used when dairy producers develop formal agreements with other graziers to grow out their heifers, usually at a predetermined growth rate, until point of calving.

cost of production (COP) The summation of variable costs, cash overhead and imputed overhead costs for producing milk from the dairy enterprise.

cow comfort The general wellbeing of a cow that takes into account climatic stress, design of cow housing and stock facilities and the potential behavioural stress from herd mates and stock people. In other words, how much is she 'at peace' with her perception of the world?

cow signals, cow talk The various ways cows communicate with each other and with farm staff to indicate their wellbeing.

crude protein A crude measure of the total protein in a feed, calculated as the total nitrogen content multiplied by 6.25. It includes true protein, which provides the amino acids for animal use, and also non-protein nitrogen, such as urea.

cubicle A place for a single cow to stand or lie in, which is separated from other cubicles by walls or dividers. The cows are not tied in the cubicle and can enter and leave at will. It is also called a stall.

Dairy Smallholder Innovative Program (DSIP) The name given to a structured regional dairy development program in tropical Asia.

demand The amounts of a product or service that consumers wish to buy at a range of prices.

depreciation The loss in value of capital items as they get used or become older.

digestibility The proportion of the dry matter in a feed that gets digested; it is the difference between what is eaten and what comes out as manure.

dry matter (DM) The proportion of a feed remaining after being dried at 80 to 100°C for 24 hours or until a constant dry weight is achieved. The nutritive value and the livestock requirements of feeds are usually expressed on a dry matter, rather than a fresh weight basis.

E. coli Bacteria causing scours in calves.

EBIT Earnings Before Interest and Tax. Farm revenues less farm expenses before the payment of interest on loans and income tax.

Economic Farm Surplus (EFS) Also known as Net Farm Income or operating profit.

efficiency The ratio of the quantity of output to the quantity of input. Can be technical or economic.

energy The part of a feed that is used as 'fuel' in carrying out the cow's bodily functions.

enterprise A farming activity or the production of a particular commodity or group of related commodities.

equity The value of an asset after liabilities have been accounted for. Also called net worth. What the business owes the owners.

expense Payment (in cash) plus imputed labour depreciation. Also called expenditure.

farm blindness This occurs when farmers think that what they see every day on their own farm is normal.

Farm business management (FBM) This is the process by which farm costs and returns are calculated to derive farm profit.

feed conversion ratio (FCR) This is a measure of the efficiency of conversion of feed into milk.

feedlot The farming system in which stock are fed entirely in a shed with no access to pastures.

feedpad That part of the pasture-based farm that is used to supplement feeding of dairy stock on an area formed, surface or stocked at a rate that precludes vegetation.

fibre The cell wall, or structural material, in a plant made up of (among other things) cellulose, hemicellulose, and lignin.

financial statement An annual statement summarising a business's financial activities over the last 12 months.

Food and Agriculture Organization (FAO) The organisation within the United Nations structure that documents agricultural statistics and facilitates development in agriculture, particularly in the poorer countries of the world.

foot and mouth disease (FMD) This is a highly contagious and often fatal disease affecting cattle.

free stalls Cubicles specifically designed for individual stock to rest in.

Golden Rules The name given to eight sets of best management practices for smallholder dairy farmers.

Gross Farm Income (GFI) The sum of total farm cash income plus changes in stock inventory.

growth Increase in net worth (wealth) over time. Measured as change in equity.

herringbone The name given to a type of milking parlour design where cows can be individually handled for milk harvesting.

hierarchy An ordered sequence of individuals or groups of individuals in a social system which is based upon some ability or characteristic, most often to act aggressively towards or displace group members or to have priority of access to some areas of the cowshed.

imputed Same as unpaid or book value.

input Any resource used in production, such as land, labour, capital or purchased goods or services.

interest A charge made for borrowing a sum of money or due on late payment of an account.

Key Performance Indicator (KPI) A numerical descriptor of some aspect of herd management or farm performance that can be used as a realistic target for future improved farm management programs.

lactation anoestrus This occurs when high genetic merit cows utilise excess levels of their body reserves to maintain milk yields, hence will rapidly lose weight and upset the hormonal balances to allow them to regain their normal oestrous cycle soon after calving.

liability The amount of money borrowed from someone else. Also called borrowings and debt.

live weight at first calving (LWFC) An important measure of the success of young stock feeding and management.

loose housing A shed in which cows are not tied up.

maintenance requirement The energy needed for essential body functions, such as blood circulation, breathing, keeping warm or cool, digestion and tissue repair.

management The process of setting and achieving (profitable) goals, using land, labour, management and capital.

metabolisable energy (ME) The amount of energy provided by a feed after deducting energy lost to faeces, urine, heat and gas production; it is the energy available to be used by the animal for its metabolic activities.

Methylene Blue Reductase Test (MBRT) The name of a test to quantify the bacterial contamination of raw milk post-harvesting.

milk gross income (MGM) Milk revenue less direct milk production expenses.

milk income less feed costs (MIFC) A measure of profitability calculated from the income from milk sales less the total cost of feed inputs. It is also called feeding profit.

net farm income (NFI) The gross farm income less variable and overhead costs. It is also called Operating Profit, Earnings Before Interest and Tax (EBIT) and Economic Farm Surplus.

net profit The operating profit less the financial costs.

neutral detergent fibre (NDF) A measure of all the fibre (hemicellulose, lignin, and cellulose) in a feed; it indicates how bulky the feed is.

non-protein nitrogen (NPN) Not actually protein but simple nitrogen; however, microbes can make protein from simple nitrogen if enough energy (carbohydrates) is available in the rumen at the same time.

operating profit The financial profit derived from all the capital invested on the farm. Also called the Net Farm Income.

overhead cost An expense that cannot be directly attributed to the production and/or harvesting of produce for sale. Also defined as a cost that remains the same when the level of input or output changes, or one that must be met and is not affected by the amount of produced in a year. Also called fixed cost.

partial mixed ration (PMR) The supplement fed to grazing stock, usually prepared in a mixer wagon.

partitioning The metabolic division of energy intake (above the maintenance requirement) between live weight gain and milk production.

per In each or for each.

peri-urban farm Farms located on the outskirts of large towns and cities, benefiting from close proximity to consumers – hence markets – as well as being able to easily source forages in nearby rural areas and agro-industrial by-products from urban food processors.

pH A measure of acidity or alkalinity on a scale from 1 (extremely acid) to 14 (extremely alkaline).

position description (PD) The description of a particular staff position, usually on a large dairy farm.

profit A general term indicating some kind of surplus from the year's farming operations.

profitability The ratio of the value of goods produced against the value of the resources used in their production. Profitability is different from profit. Profit is an amount of money; profitability is that amount compared to the resource used to produce that profit.

Profitable Dairy Systems (PDS) The international dairy farm consulting business owned by the senior author.

pros and cons The arguments in favour of and against.

public private partnership (PPP) The combination of government (public) and commercial (private) agreements in a dairy development project.

quality assurance (QA) A structured set of best management practices.

quality In relation to feeds, it is an indication of the level of energy and digestibility. In relation to milk, it refers to the level of various contaminants in milk, such as bacterial, chemical or any other adulterations that can be detected.

resource A factor of production, commonly classified under labour, land, capital or raw materials.

Return On Assets (ROA) Profit before interest, as a percentage of the value of the total farm assets.

Return On Equity (ROE) Profit, after interest expenses, compared to equity.

revenue Receipts (cash) plus imputed gain in asset value.

risk A situation with uncertain outcomes, but a case where some probabilities can be formed about the outcome. This is in contrast to uncertainty where no probabilities can be formed about uncertain events happening.

Rp Rupiah, the Indonesian unit of money.

service providers The farmer's network of people who provide services, equipment and 'good ideas' to improve their farm performance, profits, and hence long-term sustainability.

SHD Smallholder dairy.

slatted floor A combination of solid parts (slats), which would support the lower surface of the claw of the cow, and gaps (slots) which would allow manure and other liquids to pass through. (Also called slotted floor.)

Smallholder Dairy Development Program (SDDP) The name given to an FAO dairy development program in Thailand, Bangladesh and Myanmar.

SNF Solids-not-fat.

solid floor A continuous flat surface which might be made of various materials and which allows full contact with and support to the lower surface of the claw of the cow.

solids-not-fat (SNF) An alternative measure of milk composition to milk protein. SNF contains milk protein, lactose and minerals, allowing milk protein % to be calculated as (SNF% – 5.4).

somatic cell count (SCC) A measure of the bacterial contamination in raw milk pre harvesting.

South East (SE) That portion of Asia to which this book mainly relates.

Standard Operating Procedures (SOP) A set of instructions for any activity or set of tasks undertaken on the farm.

strategic Long term.

stress An environmental effect on an individual that over-taxes its control systems and reduces its performance and survivability or has the potential to do so.

submission rate The proportion of the herd inseminated at least once in a given period of time (e.g. the first 10, 21, 24 or 30 days of mating).

supplement A feed or product added to the animal's diet to increase the intake of some dietary component, such as energy, protein, fibre, vitamins or minerals.

SWOT An assessment of an activity in terms of its strengths, weaknesses, opportunity and threats.

Temperature Humidity Index (THI) A system for quantifying heat stress based on temperature and humidity. The higher the index, the greater the discomfort, and this occurs at lower temperatures for higher humidities.

tie stalls A stock housing system in which stock are continually tied up.

total digestible nutrients (TDN) A system of describing dietary energy based on proximate analyses (ash, nitrogen, ether extract, and crude fibre). Formula uses crude protein, crude fibre, ether extract, and nitrogen-free extract.

total mixed ration (TMR) A complete ration for livestock thoroughly mixed then delivered to stock from a wagon.

total plate count (TPC) A measure of bacterial contamination of raw milk in millions of bacterial colony-forming units per ml of milk.

total solids (TS) or total dissolved solids (TDS) A measure of milk composition expressed in % of total milk solids (milk fat, milk protein, lactose and minerals).

tunnel ventilation barn An enclosed shed in which rapid air movement results from the negative pressure resulting from a wall full of large extractor fans pushing the air away from the shed.

UHT Ultra high temperature treated milk.

uncertainty A situation with uncertain outcomes where no probabilities can be formed about uncertain events happening. Chances are that the actual result will be different to the planned result. Volatility increases uncertainty, hence risk.

variable costs Costs that change directly according to the amount of output of the activity. Also known as direct cost, such as fuel, seed, fodder.

wastage rate A measure of losses in replacement heifers between birth and second calving.

withholding period The number of days following drug administration before milk or meat can be sold from treated animals.

References and further reading

Ahuja V, Staal S (2013) *Dairy in South East Asia: A Rapid Overview of Key Trends and Some Issues*. Prepared for World Milk Day, 1 Jun 2013. APHCA, Bangkok, Thailand.

Anggraeni A, Rowlinson P (2005a) 'Productivity of Holstein dairy cows maintained in two different conditions in Central Java, Indonesia, 1: Evaluation of milk performance'. In *Integrating Livestock-Crop Systems to Meet the Challenges of Globalisation*, Vol 2, p. T128. (Eds P Rowlinson, G Wachirapakorn, P Pakdee and M Wanapat). *Proceedings of AHAT/BSAS International Conference*, Nov 2005, Khon Kaen University, Thailand.

Anggraeni A, Rowlinson P (2005b) 'Productivity of Holstein dairy cows maintained in two different conditions in Central Java, Indonesia, 2: Environmental effects on reproductive performance'. In *Integrating Livestock-crop Systems to Meet the Challenges of Globalisation*, Vol 2, p. P106. (Eds P Rowlinson, G Wachirapakorn, P Pakdee and M Wanapat). *Proceedings of AHAT/BSAS International Conference*, Nov 2005. Khon Kaen University, Thailand.

Animal Production and Health Commission for Asia and the Pacific (APHCA) (2013) *Asia Dairy Network*. <http://www.dairyasia.org>.

Anon (2005) *Indonesia's Dairy Farming Industry SWOT Analysis 2005*. Sept 2005. Stanton, Emms and Sia, Singapore.

Asia Dairy Network (2015) *Asia Dairy Network Website Newsletters*. <http://www.dairyasia.org>.

Bos AP, Cornelissen JMR, Koerkamp PWG (2009) *Cow Power. Designs for System Innovation*. Wageningen, Lelystad, Netherlands.

Burrell DE, Moran JB (2004) 'Developing a strategic plan for Indonesia's smallholder dairy industry'. Vol 1, pp. 143–146. (Eds HK Wong, JB Liang, ZA Jelan, YW Ho, YM Goh, JM Panandam and WZ Mohamad). In *Proceedings of the 11th Animal Science Congress*, Sept 2004. Asian-Australasian Association of Animal Production Societies, Malaysia.

Chamberlain P, Moran J, Morey P, Sharma K (2015) 'The provision of dairy development consultancy services, India'. Report to Australian Trade Commission.

Chantalakhana C, Skunmun P (2002) *Sustainable Smallholder Animal Systems in the Tropics*. Kasetsart University Press, Bangkok.

Cook BG, Pengelly BC, Brown SD, Donnelly JL, Eagles DA, Franco MA, Hanson J, Mullen BF, Partridge IJ, Peters M, Schultze-Kraft R (2005) *Tropical Forages: An Interactive Selection Tool*. [CD-ROM], CSIRO, DPI&F(Qld), CIAT and ILRI, Brisbane, Australia. <http://www.tropicalforages.info>.

Delgardo C, Rosegrant M, Wada N (2003) 'Meating and milking global demand: Stakes for small-scale farmers in developing countries', pp. 13–23. In *The Livestock Revolution. A Pathway to Poverty?* ATSE Crawford Fund Conference, Canberra, Australia.

Devendra C (2001) Smallholder dairy production systems in developing countries: Characteristics, potential and opportunities for improvement. *Asian-Australasian Journal of Animal Science* **14**, 104–115.

Dugdill B, Morgan N (2008) Smallholder dairying in the Asia–Pacific region. Synthesis of nine country lessons learned studies. pp. 6–12. In *Developing an Asian Regional Strategy for Sustainable Smallholder Dairy Development. Proceedings of an FAO/APHCA/CFC Funded Workshop*, Feb 2008, Chiang Mai, Thailand. <http://www.aphca.org/publications/files/Proceedings%20of%20the%20workshop%20final_with_coverpage_revised.pdf>.

Falvey L, Chantalakhana C (1999) *Smallholder Dairying in the Tropics.* ILRI, Kenya.

FAOSTAT (2013) *FAO Statistics Online.* <http://faostat3.fao.org/browse/Q/QL/E>.

Gooch CA, Timmons MB (2005) *Tunnel Ventilation for Freestall Barns.* College of Agriculture and Life Sciences, Cornell University. <www.mnproject.org/resourcecenter/tunnelvent.pdf>.

Hemme T, Otto J (2010) *Status and Prospects for Smallholder Milk Production: A Global Perspective.* FAO, Rome.

Hooten N (2008) Dairy development for the resource poor. Lessons for policy and planning strategies. In Developing an Asian Regional Strategy for Sustainable Smallholder Dairy Development. pp 48–51, *Proceedings of an FAO/APHCA/CFC funded workshop*, Feb 2008, Chiang Mai, Thailand. <http://www.aphca.org/publications/files/Proceedings%20of%20the%20workshop%20final_with_coverpage_revised.pdf>.

Hulsen J (2011) *Cow Signals. A Practical Guide for Dairy Farm Management.* Rood Bont Publishers, Zutphen, Netherlands.

Hulsen J (2013) *Cow Signals Checkbook. Working on Health, Production and Welfare.* Rood Bont Publishers, Zutphen, Netherlands.

Koonawootrittriron S (2016) Dairy genetic improvement in Asia (Thailand). In *Dairy Asia: Towards Sustainability.* Jan 2016, Proceedings of Multi-Stakeholder Meeting Held in Muak Lek, Saraburi, Thailand.

Little S (2012) *Feeding Systems Used by Australian Dairy Farmers. Grains2Milk.* Dairy Australia website. <http://www.dairyaustralia.com.au/-/media/Documents/Animals%20feed%20and%20environment/Feed%20and%20nutrition/Feeding%20Systems%20latest?Aus%20five%20main%20feeding%20systems.pdf>.

Makeham JP, Malcolm LR (1986) *The Economics of Tropical Farm Management.* Cambridge University Press, Cambridge, UK.

Moran JB (2005) *Tropical Dairy Farming: Feeding Management for Smallholder Dairy Farmers in the Humid Tropics.*

CSIRO Publishing, Melbourne, Australia. <http://www.publish.csiro.au/nid/197/issue/3363.htm>.

Moran JB (2009a) *Business Management for Tropical Dairy Farmers.* CSIRO Publishing, Melbourne, Australia. <http://www.publish.csiro.au/nid/220/issue/5522.htm>.

Moran J (2009b) Key performance indicator's to diagnose poor farm performance and profitability of smallholder dairy farmers in Asia. *Asian-Australasian Journal of Animal Science* **22**, 1709–1717.

Moran J (2011) Factors affecting high mortality rates of dairy replacement calves and heifers in the tropics and strategies for their reduction. *Asian-Australasian Journal of Animal Science* **24**, 1318–1328.

Moran JB (2012a) *Managing High Grade Dairy Cows in the Tropics.* CSIRO Publishing, Melbourne, Australia. <http://www.publish.csiro.au/nid/220/issue/6812.htm>.

Moran JB (2012b) *Rearing Young Stock on Tropical Dairy Farms in Asia.* CSIRO Publishing, Melbourne, Australia. <http://www.publish.csiro.au/nid/220/issue/6810.htm>.

Moran J (2013) Addressing the key constraints to increasing milk production from smallholder dairy farms in tropical Asia. *International Journal of Agriculture and Biosciences* **2**(3), 90–98.

Moran J (2014) *The Feeding of By-products on Smallholder Dairy Farms in Asia and Other Tropical Regions: Final Report on E-Conference Conducted in late 2013.* Asia Dairy Network. <http://dairyasia.org/>.

Moran J (2015) New insights into categorizing the status of milking cows on any dairy farm. *Broadening Horizons, Feedipedia – Animal Feed Resources Information System,* No 16, Jun 2015. <http://www.feedipedia.org/sites/default/files/public/BH_016.pdf>.

Moran J (2017) 'Improving smallholder dairy farming in tropical Asia'. In *Achieving Sustainable Production of Milk,* Vol. 2: Safety, Quality and Sustainability. (Ed. N van Belzen). Burleigh Dodds Science Publishing, Cambridge, UK.

Moran J, Brouwer J (2012) Planning dairy development programs in Asia. In *Proceedings of 15th Congress of Asian Australasian Association of Animal Production,* Nov 2012, Bangkok, Thailand.

Moran J, Brouwer J (2013a) Interrelationships between measures of cow and herd performance and farm profitability on Malaysian dairy farms. *International Journal of Agriculture and Biosciences* **2**(5), 221–233.

Moran J, Brouwer J (2013b) Feeding management and farmer concerns about the constraints to production on Malaysian dairy farms. *International Journal of Agriculture and Biosciences* **2**(5), 234–241.

Moran J, Doyle R (2015) *Cow Talk: Understanding Dairy Cow Behaviour to Improve Their Welfare on Asian Farms.* CSIRO Publishing, Melbourne, Australia. <http://www.publish.csiro.au/nid/220/issue/7929.htm>.

Moran JB, Morey P (2015) Increasing the production of raw milk in Indonesia and other South-East Asian countries. *Proceedings of 6th International Seminar on Tropical Animal Production (ISTAP)*, Oct 2015, Yogyakarta, Java, Indonesia.

Moran J, Bernard J, Young A (2016) Prioritising improvements to traditional management practices on small holder dairy farms in the humid tropics of Asia. *International Journal of Agriculture and Biosciences* **5**(2), 73–81.

Morey P (2011) *Dairy Industry Development in Indonesia. Market Report for International Finance Corporation.* <http://www.ifc.org/wps/wcm/connect/93f48d00470e3bf883ffd7b2572104ea/Dairy+Industry+Development-2011.pdf?MOD=AJPERES>.

Seruni AP, Utami HD, Nugroho BA (2015) *The Insight into Financial Performance of Commercial and Small-scale Dairy Farming at Thailand.* <http://fapet.ub.ac.id/wp-content/uploads/2015/01/THE-INSIGHT-INTO-FINANCIAL-PERFORMANCE-OF-COMMERCIAL-AND-SMALL-SCALE-DAIRY-FARMING-AT-THAILAND_Ainun-Pizar-Seruni-115050107111024.pdf>.

Staal SS, Pratt AN, Jabbar M (2008) 'Dairy development for the resource poor. 1. A comparison of dairy policies and development in South Asia and East Africa'. Pro-Poor Livestock Policy Initiative Working Paper No 44–1, FAO, Rome.

STOAS (1999) Fertilisation of pasture. pp. 69–71. In *Modern Dairy Farming in Tropical and Subtropical Regions.* Vol 1. STOAS, Wageningen, Netherlands.

Young A (2015) Whole farm problems with heat stress. It's not just the lactating cows. pp 51–57, *Proceedings 6th International Seminar on Tropical Animal Production*, Oct 2015, Yogyakarta, Java, Indonesia.

Young D (2008) Developing the Asia smallholder dairy development strategy (DDS). pp. 67–70. In *Developing an Asian regional strategy for sustainable smallholder dairy development. Proceedings of an FAO/APHCA/CFC Funded Workshop*, Feb 2008. Chiang Mai, Thailand. <http://www.aphca.org/publications/files/Proceedings%20of%20the%20workshop%20final_with_coverpage_revised.pdf>.

Index

www.ingramcontent.com/pod-product-compliance
Ingram Content Group UK Ltd.
Pitfield, Milton Keynes, MK11 3LW, UK
UKHW051449290926
15471UKWH00008B/1109

9 781486 306466